Environmental Engineering for the 21st Century: Addressing Grand Challenges

COMMITTEE ON THE GRAND CHALLENGES AND
OPPORTUNITIES IN ENVIRONMENTAL ENGINEERING
FOR THE TWENTY-FIRST CENTURY

Division on Earth and Life Studies
Division on Engineering and Physical Sciences
National Academy of Engineering

Board on Agriculture and Natural Resources
Board on Atmospheric Sciences and Climate
Board on Chemical Sciences and Technology
Board on Earth Sciences and Resources
Board on Energy and Environmental Systems (DEPS)
Board on Environmental Studies and Toxicology
Board on Life Sciences
NAE Office of Programs
Ocean Studies Board
Water Science and Technology Board

A Consensus Study Report of

The National Academies of
SCIENCES · ENGINEERING · MEDICINE

THE NATIONAL ACADEMIES PRESS
Washington, DC
www.nap.edu

THE NATIONAL ACADEMIES PRESS
500 Fifth Street, NW
Washington, DC 20001

This activity and material is based upon work supported by the National Science Foundation under Grant No. 10002678, the U.S. Department of Energy, Office of Biological and Environmental Research, under Award No. DE-SC0016218, and the California Delta Stewardship Council under California State Award No. 1725. Any opinions, findings, conclusions, or recommendations expressed in this publication do not necessarily reflect the views of any organization or agency that provided support for the project.

International Standard Book Number-13: 978-0-309-47652-2
International Standard Book Number-10: 0-309-47652-6
https://doi.org/10.17226/25121
Cover photo credit: Prashanth Vishwanathan/IMWI

Additional copies of this report are available for sale from the National Academies Press, 500 Fifth Street, NW, Keck 360, Washington, DC 20001; (800) 624-6242 or (202) 334-3313; http://www.nap.edu/.

Copyright 2019 by the National Academy of Sciences. All rights reserved.

Printed in the United States of America

Suggested citation: National Academies of Sciences, Engineering, and Medicine. 2019. *Environmental Engineering for the 21st Century: Addressing Grand Challenges*. Washington, DC: The National Academies Press. doi: https://doi.org/10.17226/25121.

The National Academies of
SCIENCES · ENGINEERING · MEDICINE

The **National Academy of Sciences (NAS)** was established in 1863 by an Act of Congress, signed by President Lincoln, as a private, nongovernmental institution to advise the nation on issues related to science and technology. Members are elected by their peers for outstanding contributions to research. Dr. Marcia McNutt is president.

The **National Academy of Engineering (NAE)** was established in 1964 under the charter of the National Academy of Sciences to bring the practices of engineering to advising the nation. Members are elected by their peers for extraordinary contributions to engineering. Dr. C. D. Mote, Jr., is president.

The National Academy of Medicine (NAM) formerly the Institute of Medicine) was established in 1970 under the charter of the National Academy of Sciences to advise the nation on medical and health issues. Members are elected by their peers for distinguished contributions to medicine and health. Dr. Victor J. Dzau is president.

The three Academies work together as the **National Academies of Sciences, Engineering, and Medicine** to provide independent, objective analysis and advice to the nation and conduct other activities to solve complex problems and inform public policy decisions. The National Academies also encourage education and research, recognize outstanding contributions to knowledge, and increase public understanding in matters of science, engineering, and medicine.

Learn more about the National Academies of Sciences, Engineering, and Medicine at **www.nationalacademies.org.**

Consensus Study Reports published by the National Academies of Sciences, Engineering, and Medicine document the evidence-based consensus on the study's statement of task by an authoring committee of experts. Reports typically include findings, conclusions, and recommendations based on information gathered by the committee and committee deliberations. Each report has been subjected to a rigorous and independent peer review process and it represents the position of the National Academies on the statement of task.

Proceedings published by the National Academies of Sciences, Engineering, and Medicine chronicle the presentations and discussions at a workshop, symposium, or other event convened by the National Academies. The statements and opinions contained in proceedings are those of the participants and not endorsed by other participants, the planning committee, or the National Academies.

For information about other products and activities of the National Academies, please visit nationalacademies.org/whatwedo.

Environmental Engineering for the 21st Century: Addressing Grand Challenges

COMMITTEE ON GRAND CHALLENGES AND OPPORTUNITIES IN ENVIRONMENTAL ENGINEERING FOR THE TWENTY-FIRST CENTURY

Domenico Grasso, *Chair,* University of Michigan, Dearborn
Craig H. Benson (NAE), University of Virginia, Charlottesville
Amanda Carrico, University of Colorado, Boulder
Kartik Chandran, Columbia University, New York City
G. Wayne Clough (NAE), Georgia Institute of Technology, Atlanta
John C. Crittenden (NAE), Georgia Institute of Technology, Atlanta
Daniel S. Greenbaum, Health Effects Institute, Boston, MA
Steven P. Hamburg, Environmental Defense Fund, Belmont, MA
Thomas C. Harmon, University of California, Merced
James M. Hughes (NAM), Emory University, Atlanta, GA
Kimberly L. Jones, Howard University, Washington DC
Linsey C. Marr, Virginia Polytechnic Institute and State University, Blacksburg
Robert Perciasepe, Center for Climate and Energy Solutions, Arlington, VA
Stephen Polasky (NAS), University of Minnesota, St. Paul
Maxine L. Savitz (NAE), Honeywell, Inc. (*retired*), Los Angeles, CA
Norman R. Scott (NAE), Cornell University, Ithaca, NY
R. Rhodes Trussell (NAE), Trussell Technologies, Inc., Pasadena, CA
Julie B. Zimmerman, Yale University, New Haven, CT

NATIONAL ACADEMIES OF SCIENCES, ENGINEERING, AND MEDICINE STAFF

Stephanie E. Johnson, Study Director, Water Science and Technology Board
Nancy Huddleston, Communications Director, Division on Earth and Life Studies
Kara Laney, Senior Program Officer, Board on Agriculture and Natural Resources
Anne Johnson, Consultant Science Writer
Brendan R. McGovern, Research Assistant, Water Science and Technology Board

Environmental Engineering for the 21st Century: Addressing Grand Challenges

PREFACE

The National Academies of Sciences, Engineering, and Medicine convened a committee of prominent environmental engineers, scientists, and policy experts to identify grand challenges and opportunities in environmental engineering for the next several decades. The committee was also asked to describe how the field of environmental engineering and its aligned sciences might evolve to better address these needs. The study was sponsored by the National Science Foundation, the Department of Energy, and the Delta Stewardship Council (see Appendix A for the full statement of task).

Rather than focusing on specific environmental engineering challenges, the committee chose to identify the most pressing challenges of the 21st century for which the expertise of environmental engineering will be needed to help resolve or manage. The committee sought input from the scientific community, nongovernmental organizations, and the broader public and benefited from ideas produced from four prior Association of Environmental Engineering & Science Professors workshops on Grand Challenges.[1] In total, over 450 ideas for grand challenges were submitted. This report identifies five broad and interconnected challenges that need to be addressed to ensure that people and ecosystems thrive. For each challenge, the committee discussed areas where knowledge and technological advances are needed and provided examples of potential roles for environmental engineers.

The study is modeled on the *NAE Grand Challenges for Engineering*, a 2008 study from the National Academy of Engineering (NAE) that identified 14 engineering challenges that, if achieved, have the potential to radically improve life on the planet. The NAE Grand Challenges cover health, sustainability, security, and joy of living, and several overlap with the challenges discussed here, including to provide access to clean water, develop carbon sequestration methods, make solar energy affordable, manage the nitrogen cycle, and restore and improve urban infrastructure. The NAE study and subsequent outreach efforts have inspired numerous educational initiatives, including the undergraduate NAE Grand Challenges Scholars Program aimed at creating engineers specially equipped to address 21st century challenges. The committee hopes that this study will help produce substantive progress toward meeting the critical challenges of the 21st century through advances in environmental engineering education, research, and practice.

This Consensus Study Report was reviewed in draft form by individuals chosen for their diverse perspectives and technical expertise. The purpose of this independent review is to provide candid and critical comments that will assist the National Academies of Sciences, Engineering, and Medicine in making each published report as sound as possible and to ensure that it meets the institutional standards for quality, objectivity, evidence, and responsiveness to the study charge. The review comments and draft manuscript remain confidential to protect the integrity of the deliberative process. We thank the following individuals for their review of this report: Robert F. Breiman, NAM, Emory University; Paul R. Brown, Paul Redvers Brown Inc; Virginia Burkett, U.S. Geological Survey; Greg Characklis, University of North Carolina; Paul Ferrão, Technical University of Lisbon, Portugal; Peter Gleick, NAS, Pacific Institute for Studies in Development, Environment, and Security; Patricia Holden, University of California, Santa Barbara; James H. Johnson Jr., Howard University; Michael C. Kavanaugh, NAE, Geosyntec Consultants; Daniele Lantagne, Tufts University; David Lobell, Stanford University, Al McGartland, U.S. Environmental Protection Agency; James R. Mihelcic, University of South Florida; Patrick M. Reed, Cornell University; Jerry L. Schnoor, NAE, University of Iowa; Peter Schultz, ICF International; John Volckens, Colorado State University; Robyn S. Wilson, Ohio State University; and Yannis C. Yortsos, NAE, University of Southern California, Los Angeles.

Although the reviewers listed above provided many constructive comments and suggestions, they were not asked to endorse the conclusions or recommendations of this report nor did they see the final draft before its release. The review of this report was overseen by Chris Hendrickson, Carnegie Mellon University, and Jared Cohon, Carnegie Mellon University. They were responsible for making certain that an independent examination of this report was carried out in accordance with the standards of the National Academies and that all review comments were carefully considered. Responsibility for the final content rests entirely with the authoring committee and the National Academies.

---

[1] See https://aeesp.org/nsf-aeesp-grand-challenges-workshops.

Environmental Engineering for the 21st Century: Addressing Grand Challenges

CONTENTS

Introduction / **1**

Grand Challenge 1: Sustainably Supply Food, Water, and Energy / **8**

Grand Challenge 2: Curb Climate Change and Adapt to Its Impacts / **26**

Grand Challenge 3: Design a Future Without Pollution or Waste / **44**

Grand Challenge 4: Create Efficient, Healthy, Resilient Cities / **54**

Grand Challenge 5: Foster Informed Decisions and Actions / **66**

The Ultimate Challenge for Environmental Engineering:
   Preparing the Field to Address a New Future / **78**

Endnotes and Figure Sources / **90**

APPENDICES

A Statement of Task / **101**

B Biographical Sketches of Committee Members / **102**

Environmental Engineering for the 21st Century: Addressing Grand Challenges

EXECUTIVE SUMMARY

Environmental engineers support the well-being of people and the planet in areas where the two intersect. Over the decades the field has improved countless lives through innovative systems for delivering water, treating waste, and preventing and remediating pollution in air, water, and soil. These achievements are a testament to the multidisciplinary, pragmatic, systems-oriented approach that characterizes environmental engineering.

The future holds daunting challenges for human society and our environment. Populations are expanding, demand for resources is increasing, the climate is changing, and humanity's impacts on the planet continue to mount. Will we be able to achieve a better quality of life for our growing population without compromising the ability of future generations to achieve the same?

This study, authored by 18 of the nation's leading environmental engineers, scientists, and policy experts under the auspices of the National Academies of Sciences, Engineering, and Medicine, outlines the crucial role for environmental engineers in this period of dramatic growth and change. The report identifies five pressing challenges of the 21st century that environmental engineers are uniquely poised to help advance:

1: **Sustainably supply food, water, and energy**
2: **Curb climate change and adapt to its impacts**
3: **Design a future without pollution and waste**
4: **Create efficient, healthy, resilient cities**
5: **Foster informed decisions and actions**

The report's vision is ambitious. The challenges ahead are substantial. Yet every day, environmental engineers are making progress, both by applying existing knowledge and skills and by advancing research and innovation to generate new insights and achievements. By refocusing and redoubling its efforts to advance practical, impactful solutions for humanity's multifaceted, vexing problems, the field of environmental engineering can build on its past successes—and chart new territory—in the decades ahead.

INTRODUCTION

Since the dawn of civilization, humans have transformed the environment to accommodate and satisfy their needs. Advances in agriculture, mining, manufacturing, transportation, and energy production, for example, have dramatically improved standards of living over the centuries. However, this progress has been achieved at a cost to Earth's natural systems and has yet to be more equitably distributed to all. Human impacts on the environment accelerated with the advent of the Industrial Age and the subsequent rapid growth of the human population, creating significant areas of friction between human societies and the environment. At its worst, the human presence is manifest in pollution hanging over cities; sprawling development in place of forests; hazardous chemicals permeating rivers, lakes, and soil; vanishing species; and a changing climate.

The field of environmental engineering emerged to support human and environmental needs while mitigating adverse impacts associated with human activities. Propelled by public sentiment in support of protecting natural resources and human health and by laws aimed at curtailing some of the most egregious forms of environmental damage, the field has achieved remarkable successes over the past several decades. However, the solutions of the past will not be sufficient to address the problems of the future. As humanity faces mounting and diverse challenges, the field of environmental engineering must build on its unique strengths, inspire and implement visionary solutions, and continue to evolve in order to serve the best interests of people and the planet.

What Is Environmental Engineering?

Environmental engineering is best characterized by the vast array of issues that its practitioners address. Broadly, environmental engineers design systems and solutions at the interface between humans and the environment. Historically, this work focused on the provision of water and treatment of wastewater, drawing upon the field's roots in sanitation system design and public health protection. In the 1970s the term environmental engineering replaced the previous term, sanitary engineering, as the field's focus broadened to include the mitigation of pollution in air, water, and soil. Around the same time, the field's approach to design shifted from a focus on engineered treatment systems toward a greater emphasis on ecological principles and processes. More recently, the field has expanded further to address emerging contaminants, chemical exposures from goods and materials, and endeavors such as green manufacturing and sustainable urban design.

To support these activities, many environmental engineers have acquired expertise in a wide variety of domains, including hydrology, microbiology, chemistry, systems design, and civic infrastructure. About half of practicing environmental engineers have graduate degrees; practitioners apply their craft to a wide range of

areas in industry, government, nonprofits, and academia. Trained to take a systems-level approach to problems, environmental engineers often act as a bridge among scientists, other engineers, decision makers, and communities to assess options, weigh trade-offs, and design cost-effective, pragmatic solutions.

The discipline of environmental engineering has no single, widely agreed-upon definition. This report does not focus on defining the field as it is, but rather seeks to outline a vision for the ways in which environmental engineering expertise, skills, and areas of focus can help address future challenges. Fulfilling this vision will require a new model for environmental engineering practice, education, and research—building on and complementary with the field's traditional core competencies—as outlined in the report's final chapter.

New Pressures in the 21st Century

In this century, human pressure on the environment will accelerate. Life expectancy has increased substantially across the globe over the past several decades as living conditions have improved and is projected to continue to increase.[5] The United Nations predicts that by 2050 the world's population will reach roughly 9.8 billion people, an increase of approximately 30 percent from today.[6] As human

BUILDING ON A REMARKABLE LEGACY

Although the term environmental engineering has been in use for only a few decades, the field's roots reach back centuries. Romans built sophisticated sewage disposal and water supply systems, some of which still deliver water to Rome today. In the new world, the Inca and the Maya developed innovative systems to distribute clean water to great cities such as Cusco and Tikal. The beginnings of modern-day environmental engineering are typically traced to the creation of the first municipal drinking water filtration systems, the first continuously pressurized drinking water supply, and the first large-scale municipal sanitary sewer in 19th century London. These and subsequent advancements markedly improved people's quality of life by curbing the spread of disease. In the early 20th century, chlorine-based disinfection for water treatment and advances in wastewater treatment contributed to a drastic decline in urban mortality rates.[1]

Environmental engineering continued to evolve throughout the 20th century as a series of environmental crises sparked the creation of new laws aimed at preventing and mitigating pollution in air, water, and soil. After London's Great Smog of 1952 killed thousands of people, the Parliament of the United Kingdom passed the first major legislation aimed at limiting emissions from households and industries. In the United States, debilitating smog over Los Angeles and other

populations grow, so too will humanity's demand for natural resources and impacts on natural systems. These impacts will play out in different ways in different areas. At least two-thirds of the population in 2050 will live in cities, compounding pressures on urban systems that provide clean water, food, energy, and sanitation. Rapid economic and population growth in lower-income countries threatens to overwhelm basic infrastructure and drive sharp increases in pollution, just as the developed world experienced in the early 20th century. At the same time, countries of all income levels face new types of challenges—many driven by climate change—that existing policies, technologies, and infrastructures are not equipped to handle.

Most environmental engineering expertise is concentrated in developed countries, but some of the most vexing challenges are concentrated in the world's poorer regions. More than 10 percent of humanity continue to live off less than $1.90 per day and lack access to basic services and economic opportunity.[7] More than 2 billion people still lack access to basic sanitation services,[8] more than 1 billion are without electricity,[9] and more than 3 billion people rely on household energy sources that produce dangerous indoor air pollutants.[10] Unsafe air and water rank among the major contributors to disease and death worldwide.[11] Despite economic progress, meeting the basic human needs of the large swath of the world's population who live in extreme poverty will remain a monumental task in the decades ahead.

At the same time, many more people are experiencing an improved standard of living. The proportion of people living in extreme poverty has been reduced by half since 1990.[12] Recent economic growth in China, Brazil, and India has been lifting about 150 million people out of poverty and into the middle class each year.[13] Although undoubtedly positive for people's well-being and quality of life, this growth also has the potential to create or exacerbate some of the same types

---

U.S. cities from vehicle emissions led to the passage of the Clean Air Act of 1970. Environmental engineers, working with atmospheric chemists and other scientists, responded by developing models of pollution and its sources, monitoring emissions, helping ensure compliance with regulations, and designing and implementing technologies to improve air quality. Such efforts resulted in a two-thirds drop in U.S. emissions of common air pollutants between 1970 and 2017.[2]

The same period saw a major movement to reduce water pollution. After the 1969 fire on Ohio's Cuyahoga River called public attention to the widespread practice of dumping industrial and household wastes into rivers and streams, the U.S. Clean Water Act of 1972 banned the discharge of pollutants from pipes and other point sources into navigable waters without a permit. In 1974, Congress passed the Safe Drinking Water Act establishing standards for public water systems. Environmental engineers work to support the enforcement of these laws by developing water treatment technologies along with new analytical methods and modeling tools to quantify and reduce contamination of rivers and streams.

Another infamous episode focused the public and environmental engineers on contamination of soils and groundwater. More than 21,000 tons of hazardous chemicals dumped into a 70-acre industrial landfill near Love Canal, New York, during the 1950s and 1960s seeped into waterways and soil, affecting the health of hundreds of residents.[3] Responding to the disaster, Congress in 1980 passed a law launching the Superfund program, which called on the U.S. Environmental Protection Agency to develop remedial actions and treatment technologies to reduce pollutants at designated sites.[4] Environmental engineers today play a crucial role in carrying out this charge by providing technical expertise to assess and remediate existing contaminants and by designing new processes and disposal methods to prevent future contamination.

SUSTAINABLE DEVELOPMENT GOALS

A vision for responsibly improving quality of life in the world's poorer regions is embodied in the United Nations' 2030 Agenda for Sustainable Development, which articulates 17 strategic goals designed "to end poverty, protect the planet, and ensure prosperity for all."[14] While environmental quality has the potential to contribute to all of these goals, at least 10 of them relate directly or indirectly to the work of environmental engineers:

Goal 2: Zero Hunger
Goal 3: Good Health and Well-Being
Goal 6: Clean Water and Sanitation
Goal 7: Affordable and Clean Energy
Goal 9: Industry, Innovation, and Infrastructure
Goal 11: Sustainable Cities and Communities
Goal 12: Responsible Consumption and Production
Goal 13: Climate Action
Goal 14: Life Below Water
Goal 15: Life on Land

of environmental problems that wealthier countries have grappled with in the past. Some mistakes of the past may be avoided with the benefit of hindsight, public awareness, and new technology. Nonetheless, it is expected that increased purchasing power and consumption preferences of the world's growing middle class will generally lead to increases in resource and energy use, with negative implications for ecosystems, biodiversity, and human health. The United Nations' Sustainable Development Goals offer a framework to guide economic development while minimizing its potential downsides (see sidebar). The grand challenges for environmental engineers outlined in this report align closely with many of these goals.

In addition to drivers related to population growth, urbanization, poverty, and economic development, climate change adds new complexity to nearly every environmental challenge. Expected increases in extreme weather, including heat waves, drought, hurricanes, wildfires, and floods place enormous strain on water supplies, agriculture, and the built environment. Global warming is already contributing to the reemergence of pathogens and spread of insect-borne diseases to new regions. For the increasing number of people living near a coast, sea-level rise combined with storm surge has become a threat to life and property. These trends pose urgent threats in developing and developed countries alike.

A New Vision for Environmental Engineering

Environmental engineers were instrumental in pulling the United States and other countries out of the depths of environmental crises such as Love Canal and urban smog. Rivers in Ohio no longer catch fire. Cholera and other once-prevalent waterborne diseases are now so rare in the United States that lightning strikes pose a greater threat. These successes, worthy of celebration, reflect the value of the field's approach to creating systems and solutions that are grounded in sound scientific, ecological, and engineering principles while being cost-effective, feasible, and acceptable for the many stakeholders that environmental engineers serve.

But these battles are not over. Pollution and waterborne diseases persist around the globe. Rivers are still catching fire. Billions of people suffer from inadequate access to clean water, food, sanitation, and energy. As the human population continues to grow, demands intensify and humanity's mark on the planet deepens. In short, the challenges ahead are of a different nature and a larger scale than those faced in the past.

Today's environmental engineers also operate in a different policy context than the one that fueled past achievements. The types of sweeping laws that directed public attention and funding toward large-scale infrastructure expansion, basic research, and technology development for environmental remediation in the 1970s-1990s have not emerged to address today's national and global challenges. Legislation may not be the primary drivers of future innovation.

As we face this period of dramatic growth and change, it is time to step back and consider new roles that environmental engineers might play in meeting human and environmental needs. Although efforts to characterize, manage, and remediate existing environmental problems are still essential, environmental engineers must also turn their skills and knowledge toward the design, development, and communication of innovative solutions that avoid or reduce environmental problems. The core competencies of environmental engineering, which emphasize not only specific goals related to human needs and the condition of the environment but holistic consideration of the consequences of our actions, are uniquely valuable in developing the solutions that will be needed in the coming decades.

The report identifies five pressing challenges for the 21st century that environmental engineers are uniquely poised to help advance:

1: **Sustainably supply food, water, and energy**
2: **Curb climate change and adapt to its impacts**
3: **Design a future without pollution and waste**
4: **Create efficient, healthy, resilient cities**
5: **Foster informed decisions and actions**

These grand challenges stem from a vision of a future world where humans and ecosystems thrive together. Although this is unquestionably an ambitious vision, it is feasible—and imperative—to achieve significant steps toward these challenges in both the near and long term.

The challenges provide focal points for evolving environmental engineering education, research, and practice toward increased contributions and a greater impact. Implementing this new model will require modifications in the educational curriculum and creative approaches to foster interdisciplinary research on complex social and environmental problems. It will also require broader coalitions of scholars and practitioners from different disciplines and backgrounds, as well as true partnerships with communities and stakeholders. Greater collaboration with economists, policy scholars, and businesses and entrepreneurs is needed to understand and manage issues that cut across sectors. Finally, this work must be carried out with a keen awareness of the needs of people who have historically been excluded from environmental decision making, such as those who are socioeconomically disadvantaged, members of underrepresented groups, or those otherwise marginalized.

The inevitable challenges we will face over the next 30 to 50 years are daunting, but a better future is possible. By learning from the past, capitalizing on existing knowledge and skills, and growing into new roles, environmental engineers have the power to engineer a healthier and more resilient world.

INSPIRED BY ENGINEERING GRAND CHALLENGES

This report was inspired in part by the National Academy of Engineering's Grand Challenges for Engineering, announced in 2008. The effort is aimed at inspiring young engineers across the globe to address the biggest challenges facing humanity in the 21st century. An international group of leading technological thinkers identified 14 challenges within the crosscutting themes of sustainability, health, security, and joy of living. Seven of those challenges (in green) require significant input from environmental engineers.

Advance Personalized Learning	Secure Cyperspace
Make Solar Energy Economical	**Provide Access to Clean Water**
Enhance Virtual Reality	Provide Energy from Fusion
Reverse-Engineer the Brain	Prevent Nuclear Terror
Engineer Better Medicines	**Manage the Nitrogen Cycle**
Advance Health Informatics	**Develop Carbon Sequestration Methods**
Restore and Improve Urban Infrastructure	**Engineer the Tools of Scientific Discovery**

GRAND CHALLENGE 1:

Sustainably Supply Food, Water, and Energy

Providing life's essentials—food, water, and energy—for the world's growing population is a major challenge. Doing so in a manner that does not threaten the environment and the health or productivity of future generations is an even bigger challenge.

The challenges differ in high- and low-income countries. In low-income countries the infrastructure to supply water and energy and manage wastewater in many cases simply does not exist, and economic and social barriers put basic services out of reach for billions of people. Nearly 800 million people worldwide are undernourished;[15] nutrition-related factors contribute to 45 percent of deaths in children under age 5.[16] In 2015, 844 million people had no access to safe drinking water, and 2.3 billion people did not have ready access to basic sanitation services.[17] More than 1 billion people, or about 1 in 7 globally, live without electricity.[18] These issues are most severe in sub-Saharan Africa and central and southern Asia.[19] High-income countries have mature production and delivery systems to provide food, water, and energy to their populations, but these systems often waste resources and discharge harmful pollutants. In many places, water and sanitation infrastructure has outlived the planning horizon under which it was built, creating large challenges in maintaining expected water quality and reliability.

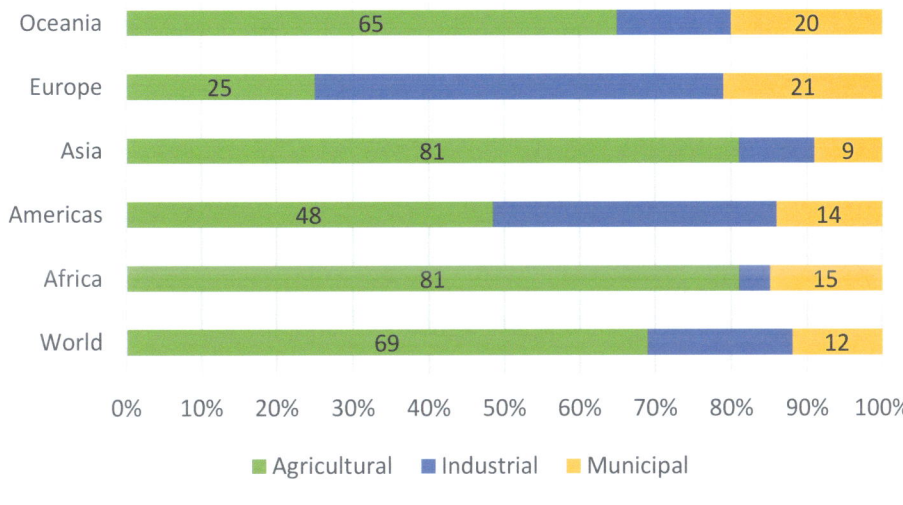

FIGURE 1-1. Water withdrawal percentages by sector and continent, 2010.

Complexities arise from the fact that food, water, and energy are inextricably linked. About 70 percent of global water withdrawals are for agricultural purposes (irrigation, livestock, or aquaculture; Figure 1-1), and agriculture represents about 80-90 percent of all consumptive use.[20] Agricultural activities release nutrients and contaminants into groundwater and downstream water bodies, degrading terrestrial and aquatic ecosystems and threatening the water resources on which humans depend.[21] The food production and supply chain is estimated to consume about 30 percent of global energy and produce about 22 percent of global greenhouse gas emissions (including landfill gas from food wastes), although there is uncertainty with such calculations.[22] The global energy mix remains dominated by fossil fuels, the extraction and the use of which involve water-intensive processing and contribute to water pollution.

In the decades ahead, sustainably supplying food, water, and energy to all will be made more difficult by population growth, increasing standards of living, and climate change. Innovation will be needed to augment supplies, improve distribution, reduce waste, increase efficiency, and reduce demand. Because the food-water-energy nexus is so tightly interwoven, potential solutions or demands in one area often have repercussions in another. A holistic, systems-oriented approach is crucial to balancing resource demands as we strive to meet the basic needs of our growing population.

Advancing Sustainable Agriculture to Feed Earth's Growing Population

Feeding a growing global population while minimizing impacts on water, soil, and climate poses substantial challenges during the next several decades.[23] By 2050, there are likely to be an additional 2.6 billion people to feed, and gains in affluence will increase energy use and the demand for water- and resource-intensive diets of meats and dairy. Climate change exacerbates pressures on water supplies and agricultural productivity[24] and increases the likelihood of disruptions in the food supply chain from storms and other factors.[25]

Almost all land area available for economically feasible food production is in use, and much of the remaining land, such as tropical forests and grassland preserves, sustains biodiversity and other important ecosystem services sustainability.[26] Increasing food supply will need to occur, not by adding new land, but by increasing efficiency and yields in existing agriculture, decreasing food waste, and changing dietary patterns.

Increasing Agricultural Yields with Reduced Environmental Impacts

Over the past century, agricultural yields have increased steadily through advances in mechanization and the use of fertilizers, pesticides, plant breeding, and irrigation technology. In the United States, such advances have ensured a safe and reliable domestic food supply while also generating a trade surplus in agricultural commodities and foods.

Advances in agricultural technologies, data collection, and computational science provide opportunities to further enhance efficiencies and increase yields. Sensors can be designed to detect and diagnose plant diseases in the field or in greenhouses to reduce lost agricultural productivity.[27] Precision applications of pesticides, herbicides, and fertilizer can dramatically reduce agrochemical use without compromising yields.[28] A better understanding of the microbiome in agriculture could improve soil structure, increase feed efficiency and nutrient availability, and boost resilience to stress and disease.[29] Selective breeding, genetic engineering, and gene editing could be used to develop crop varieties that maintain productivity under changing climate conditions.[30]

The recent explosion in the availability of data presents many opportunities to improve the resilience and efficiency of food and agricultural production. To inform decisions effectively, analysis of datasets must account for multiple factors. For example, understanding yields requires analysis of plant genetics, farm management practices, local environmental conditions, and socioeconomic factors over a range of spatial and temporal scales. Data standards and tools that can manipulate and analyze such large and complex datasets are needed to facilitate these advances.[31]

In low-income countries, some innovative efforts are improving yields and efficiency in crop production while minimizing environmental impacts. In India, for example, a new tractor-mounted seeder has been developed that allows wheat to be planted in rice paddies without burning the straw remaining after the rice harvest, a practice that simultaneously reduces air pollution by avoiding biomass burning and increases yields by retaining organic matter in the soil.[32] Advances in low-cost sensors and cell phone–based tools designed for agriculture could provide guidance to farmers on appropriate application rates of seeds, water, and fertilizer to maximize yields and prevent unnecessary inputs.

FIGURE 1-2. Wheat seeder designed to eliminate crop waste burning in rice paddies in India.

Today, some yield improvements could come at the cost of greater environmental burdens. For example, it has been estimated that it may not be possible to further increase U.S. soybean yields without sacrificing water quality and soil resources in surrounding ecosystems.[33] Environmental engineers can advance sustainable agriculture by working collaboratively with agricultural engineers and evaluating environmental benefits and impacts of innovative strategies in both low- and high-income settings.

Recent innovations in indoor aquaculture and vertical farming are expanding the possibilities of where emerging agricultural technologies can develop (see Figure 1-3). These facilities can be designed to produce food with recycled nutrients, carbon, and water to maximize water efficiency, reduce fertilizer use, and minimize pollution. Water discharged can be treated so that it is cleaner than when it enters the facility.[34] Because such farms do not require agricultural land, they can be located close to urban centers, potentially increasing resilience to supply chain interruptions and reducing the energy expended in distribution. Life-cycle analyses, considering factors such as cost, energy, water use, and pollution, will be important to developing indoor agricultural systems that are feasible and cost-efficient.

FIGURE 1-3. Using stacked growing trays, known as vertical farming, and artificial lighting, leafy greens are grown without soil, reducing water demand by 90 percent compared to conventional approaches.

Reducing Food Waste

One of the biggest opportunities to stretch the supply of food is to reduce food waste. Globally, it is estimated that as much as one-third of all food produced—1.3 billion tons per year—is lost or wasted.[35] This loss and waste occur throughout the food chain:

- In the field, when damage or spills occur during harvest or when harvesting does not occur because of economic or weather reasons;
- After harvest, when food degrades during storage;
- At the processing stage, when spills occur or food is unsuitable for processing;
- At the distribution stage, when food is damaged or degrades as it is transported or awaits sale; and
- With the consumer, when food spoils or is simply thrown away.

In lower-income countries, such as those in Latin America, Africa, and Asia, most food loss (at least 85 percent) occurs before the food reaches the consumer; in high-income countries, over 30 percent of food loss happens at the consumer level (Figure 1-4). These losses threaten food availability in food-insecure regions and represent a waste of land, energy, water, and agricultural inputs.

Technologies and systems along the entire food chain—including harvest, transportation, processing, and storage—are needed to reduce food loss from farm to plate. Nanotechnology-based protective films (in some cases edible) can lengthen shelf life, possibly without refrigeration.[36] Low-cost sensors that indicate food quality and safety could further reduce food loss. Effective strategies will also need to consider the attitudes and actions of various stakeholders that affect food waste.

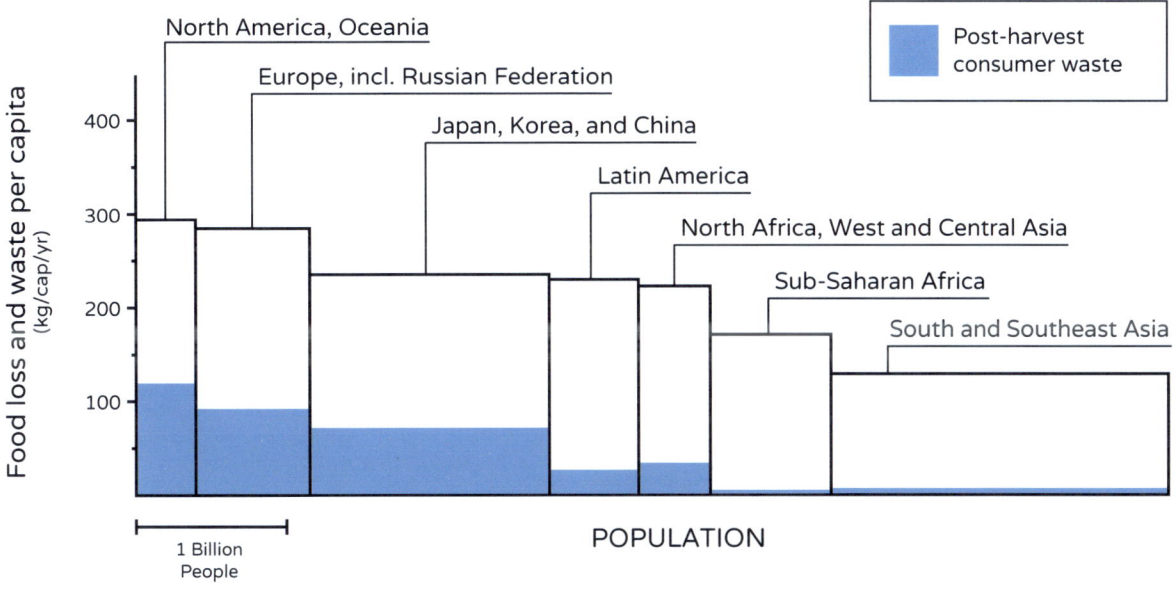

FIGURE 1-4. Food loss and waste per capita in different world regions.

Changing Dietary Patterns

Livestock farming may be responsible for as much as 14.5 percent of all human-induced greenhouse gas emissions,[37] and cattle are responsible for nearly two-thirds of these emissions. Beef and dairy farming also requires vastly more fresh water per unit of protein produced compared to plant-based equivalents. Meanwhile, it has been estimated that global meat production may grow by 12 percent between 2016 and 2026 due to population growth and increasing demand associated with rising standards of living in lower- and middle-income countries.[38]

Shifting dietary patterns to deemphasize animal-based protein, particularly beef, could reduce the environmental and resource burdens of feeding the world's population. The World Resources Institute estimates that such changes to dietary patterns could allow feeding of up to 30 percent more people with the same agricultural land and cropping patterns.[39]

A variety of meatless protein products, including innovative plant-based products and protein products grown from animal and plant tissue cells in culture, are becoming available. If such products can be produced affordably at scale and be accepted by consumers, they could reduce the demand for livestock, thereby decreasing the land, energy, and water requirements of animal-sourced protein and its associated environmental impacts while expanding food availability.

Overcoming Water Scarcity

Global water use is anticipated to increase by 55 percent by 2055, with the largest increases in Brazil, China, India, and Russia (Figure 1-5).[40] At the same time, the surface water and groundwater resources that have traditionally supplied ecosystems and human populations with fresh water are increasingly stressed. Fresh water is a limited resource, with fresh water in lakes, rivers, and groundwater comprising just 0.77 percent of the water on Earth.[41] Although

Earth's freshwater resources in total remain constant, their distribution varies widely across time and space. The beginning of the 21st century saw the Millennium drought in Australia—the worst drought recorded since European settlement.[42] California recently experienced a record-breaking multiyear drought, followed by record flooding in 2016-2017, and climate change may make such extremes more common.[43]

Water scarcity occurs when demands exceed the available water supply, leading to competition for available resources. Today, water scarcity already affects every continent and around 2.8 billion people worldwide for at least 1 month out of every year.[44] People living in water-stressed regions (Figure 1-6) are particularly vulnerable to the impact of droughts and other extreme weather events, environmental degradation, and conflict. Recently, Cape Town, South Africa, came perilously close to depleting its urban water supply. Meeting the water needs of a growing population in a manner that does not harm the environment requires innovations in water supply, increased efficiency, and strategies to reliably distribute clean water to those who need it.

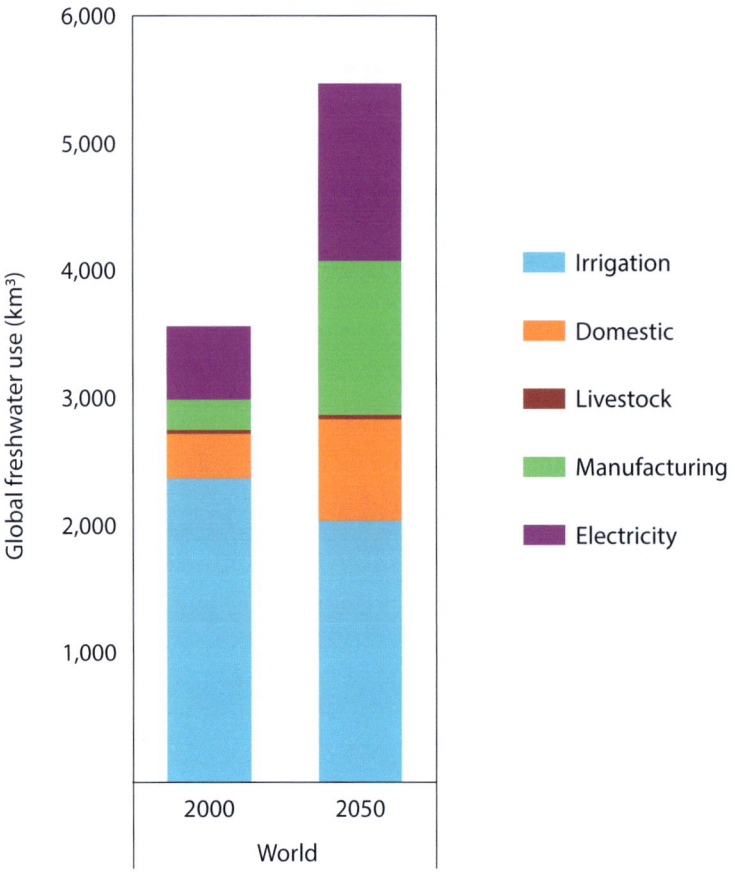

FIGURE 1-5. Global freshwater use projected for 2050, compared to the baseline in 2000. Does not include rainfed agriculture.

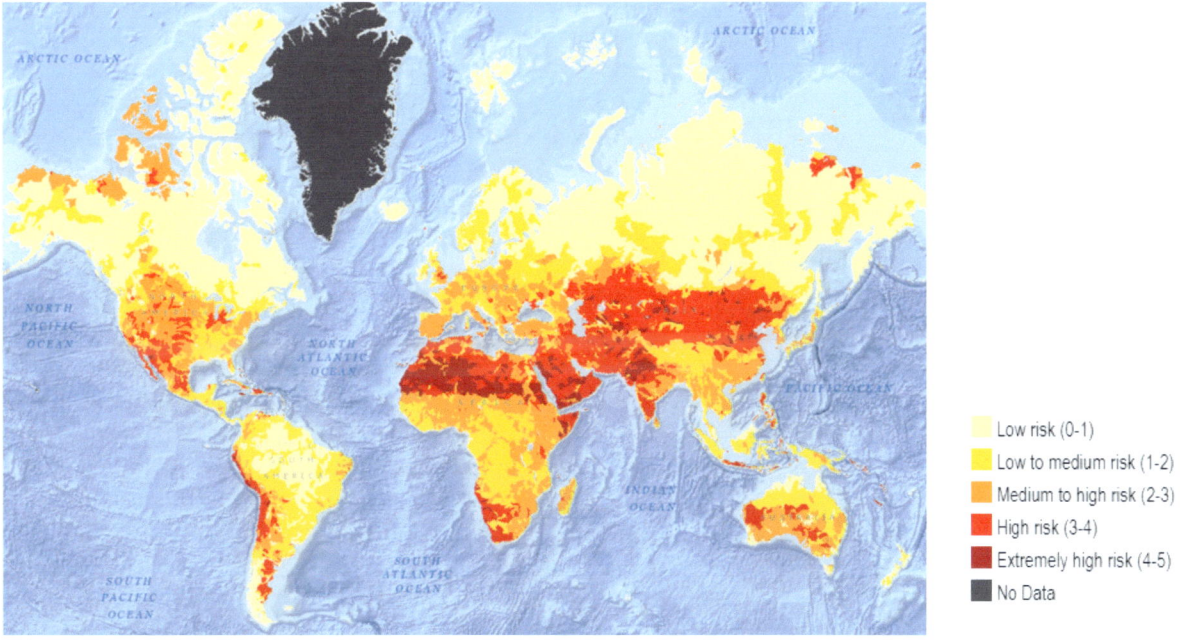

FIGURE 1-6. Map of overall water risk. Overall water risk is an aggregated measure of indicators from categories of physical risk quantity (including flood occurrence, drought severity, and upstream storage capacity), physical risk quality (including return flow ratio and upstream protected land), and regulatory and policy risk (including access to water).

Innovations in Water Supply

Fewer conventional sources of new water, such as dams and reservoirs, are being constructed in part because of increasing awareness of their environmental impacts,[45] and groundwater is being depleted worldwide at increasing rates.[46] Thus, alternative means of supplying water will be needed.

For thousands of years, people living with water scarcity have devised ways to create fresh water from seawater. As of 2015, roughly 18,000 desalination plants worldwide, almost half of them in the Middle East and North Africa, produced nearly 23 billion gallons of fresh water per day using technologies such as reverse osmosis and distillation.[47] Although important in water-scarce regions, desalination remains too expensive and energy-intensive to serve as a widespread solution for providing fresh water. Innovation and development of alternative, lower-energy approaches could change that. For example, researchers developed a membrane embedded with heat-absorbing nanoparticles that enables energy from sunlight to drive the membrane distillation process. The technology could provide off-grid desalination at the household or community scale for those who lack access to clean water.[48] Research to understand and reduce environmental impacts and to develop cost-effective approaches for brine management could also enhance the use of desalination in areas facing water scarcity.[49]

Municipalities are increasingly looking for new water supply from the recovery and reuse of water that has traditionally been simply discarded, such as stormwater, municipal wastewater, graywater (water from laundry, showers, and nonkitchen sinks), and contaminated groundwater. New technologies are making it increasingly feasible to collect stormwater or graywater at individual buildings or in neighborhoods and treat it for nonpotable uses such as irrigation, street cleaning, fire-fighting, industrial processes, heating and cooling, and toilet flushing.[50]

Cities are also turning to potable reuse systems that use advanced treatment processes to remove contaminants from wastewater to provide a drought-proof drinking water supply.[51]

Wastewater reuse is more expensive than conventional water supply alternatives such as imported water and groundwater (assuming other water alternatives are available at their traditional costs), and public acceptance of potable reuse remains a challenge. Advances are needed to reduce the cost and energy requirements of alternative supply treatment and to develop low-cost, real-time sensors for chemical and microbial contaminants (or reliable surrogates) to ensure water quality and safety.[52] The development of low-maintenance, community-scale water reuse systems with reliable quality assurance would further enhance the use of this technology.[53]

Increasing the Efficiency of Water Use

Important advances have been made over the past few decades to reduce water use. In the United States, total water withdrawals peaked in 1980, largely due to enhanced water use efficiencies from industrial production and power plant cooling,[54] although increased imports and reduced production of water-intensive goods and services, such as fruits and vegetables, may have contributed to this trend.[55] Rates of U.S. water use per person declined 40 percent between 1980 and 2010. Water is still used inefficiently in many regions, especially where it has been plentiful historically or its price has been heavily subsidized, and further advances are possible. Existing and emerging technologies and practices offer numerous opportunities to increase water use efficiency so that existing supplies can better serve the needs of a growing population and global economic growth. Agriculture is the largest water user worldwide and on every continent except Europe (see Figure 1-1). There is substantial potential worldwide for reducing water demand while maintaining or increasing agricultural output,[56] and there is already some evidence that water use efficiency strategies can improve crop quality with little

FIGURE 1-7. Small graphene sensors placed on plant leaves are used to sense water transpiration and measure plant water use so that irrigation is only applied when needed.

cost to yields.[57] Examples of water-saving techniques include farming practices, such as improved crop choice, tillage practices, and soil management, and engineering solutions, including improved precision irrigation tools and advanced ground-based sensors and remote sensing data to gauge irrigation needs more precisely (Figure 1-7).[58] Innovations are needed that enhance agricultural water productivity—the amount of crop produced per unit of water depleted (or crop per drop)—rather than simply reducing water use.[59] Current "inefficient" irrigation approaches may be recharging groundwater and supporting base flow in streams that other water users or ecosystems depend upon.

Outside of the agricultural sector, there are many other opportunities to reduce water use. Technologies to detect and prevent leaks in water distribution systems could reduce loss between the point of supply and point of use. In thermal power plants, alternative systems for cooling, such as dry cooling, could lower water demands. Technological or process improvements can help conserve water in many water-intensive industries, such as textiles, automobile manufacturing, and the beverage industry. Within homes and businesses, innovative technologies such as waterless toilets and washers could reduce water use. Innovative monitoring and communication approaches that help people understand their own water use relative to others could encourage behavioral change. Economic and policy strategies will be important, in addition to technical advances, in managing limited water supplies (see Challenge 5).

Redesigning and Revitalizing Water Distribution Systems

In high-income countries, water treatment and distribution systems developed in the early to mid 20th century led to significant improvements in public health.[60] In many locations, water infrastructure has now outlived its intended useful life, and the limits of that infrastructure are becoming evident. Older distribution system pipes are leaking and require restoration or replacement to ensure water reliability and quality.[61] In the United States, reported cases of Legionnaires' disease, caused by bacteria that can grow and spread in water systems, has increased over fourfold since 2000.[62] Some older distribution systems and many residential plumbing systems in the United States contain lead, which under certain water quality and flow conditions can become mobilized and has put residents at risk for unhealthy exposures.[63] Environmental engineers have a clear role to play in not only revitalizing and replacing these aging systems but reimagining them.

A BIG IDEA: SORTING SOLAR RADIATION TO MAXIMIZE ENERGY, FOOD, AND WATER PRODUCTION

A novel concept proposes to maximize crop production while simultaneously producing electricity and treating water supplies by unbundling the solar spectrum over a plot of land.[66] Reflective parabolic troughs can be situated above the field to collect solar energy from near-infrared and far-infrared light waves, while the solar spectrum needed for food production can pass through to the crops on the ground. The near-infrared light can be used to generate energy and the near- and far-infrared can be used to power water treatment processes through distillation or reverse osmosis. Electricity generated by the solar battery can be used for agricultural production or exported to nearby population centers. As demands for food and clean energy increase with growing populations, creative ideas such as this are needed to develop cost-effective and scalable approaches that maximize energy, food, and water supplies while reducing adverse impacts.

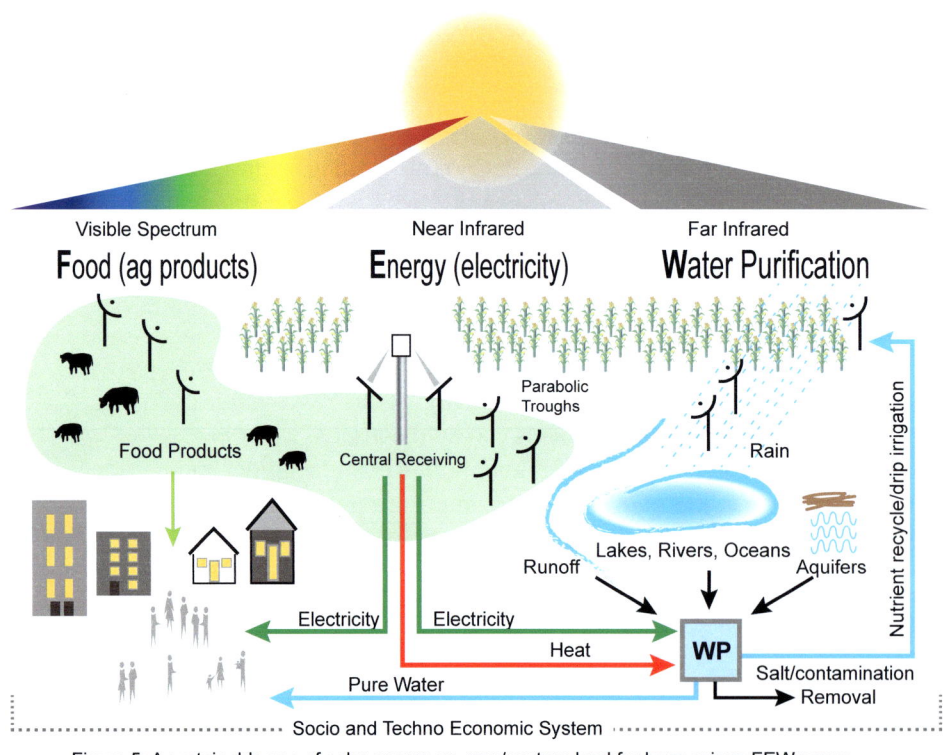

Figure 5: A sustainable use of solar energy on crop/pasture land for harmonious FEW nexus

Figure Concept of a solar spectrum unbundling in which photons are managed efficiently over crop/pasture land to simultaneously produce food, energy and water products | NOTE: WP = Water purification unit.

Low-income countries face a different set of water distribution challenges. In many regions, wastewater is discharged to surface waters without adequate treatment, polluting water bodies and denying people access to safe drinking water.[64] Advances in waterless toilets could improve access to sanitation services in low-income areas while reducing water use and pollution worldwide and enhancing the recovery of valuable resources such as energy and nutrients.[65] Where centralized infrastructure to collect, transport, and treat water and wastewater does not already exist, decentralized wastewater treatment systems using advanced technology for water reuse could enhance water supplies and recover embedded energy.

Providing Clean Energy to Meet Growing Global Demand

Access to energy is increasingly recognized as a basic human need. The UN Sustainable Development Goal 7 is to "ensure access to affordable, reliable, sustainable and modern energy for all" by 2030.[67] Improving the delivery of energy services fuels economic growth, increases productivity, and improves standards of living and health. For example, eliminating the use of unvented cookstoves that burn biomass (such as coal or dung) by supplying electricity for cooking could significantly reduce harmful indoor air pollution.

Global energy needs are expected to increase as the population grows and as more people enter the middle class. The U.S. Energy Information Administration projects that global energy consumption will grow by 28 percent between 2015 and 2040.[68] The warming of the climate is also driving changes in energy demand; it is projected that global energy demand from air conditioners will triple from 2016 to 2050, requiring new electricity capacity equivalent to the electricity capacity of the United States, the European Union, and Japan combined.[69]

Switching to More Sustainable Energy Sources

Petroleum, natural gas, and coal have been the dominant U.S. fuels for more than a century, accounting for about 80 percent of energy consumption in 2017.[70] Globally, fossil fuels also comprised about 80 percent of the primary energy supply in 2015, with nuclear and renewables such as wind, solar, hydropower, biomass, and geothermal power making up the rest.[71] Burning fossil fuels is the primary source of air pollutants as well as the greenhouse gases that drive climate change. Switching to low-carbon sources of energy and increasing energy efficiency will be essential steps to curb climate change,[72] as discussed in detail in Challenge 2.

Environmental impacts accrue not only from burning fossil fuels, but also from their production. Extraction processes, such as coal mining and drilling for oil and gas, generate air and water pollution and other land impacts that can harm local communities. For example, spills occurring during drilling processes or improperly managed mine-waste materials can contaminate surface and groundwater resources.[73] The significant amounts of water needed for unconventional natural gas extraction (hydraulic fracturing) and for cooling processes at fossil fuel electricity plants can stress local water supplies during droughts and heat waves.[74] Transportation of fuels generates additional pollution and accidents resulting in spills.[75] Continued efforts to reduce such impacts will be needed in the transition to low-carbon energy.

There are numerous ways to produce energy while emitting little or no carbon dioxide (CO_2) on an ongoing basis. In particular, solar and wind-based energy sources have gained significant traction. Other promising renewable sources that can be harnessed with minimal CO_2 emissions include hydropower from dams, tapping the energy of waves, and using geothermal energy (tapping into the heat under the Earth's surface).

Environmental impacts, costs, and benefits of renewable energy sources will need to be considered in their adoption. Wind and solar projects occupy significant amounts of land, and most wind power projects on land require service roads that add to the physical effects on the environment (Figure 1-8).[76] Siting of wind power projects atop ridgelines can disrupt scenery and recreational access. Wind turbines can kill bats and birds and harm their habitats,[77] although research on wildlife behavior has led to ways of siting and operating the turbines that help mitigate that harm.[78]

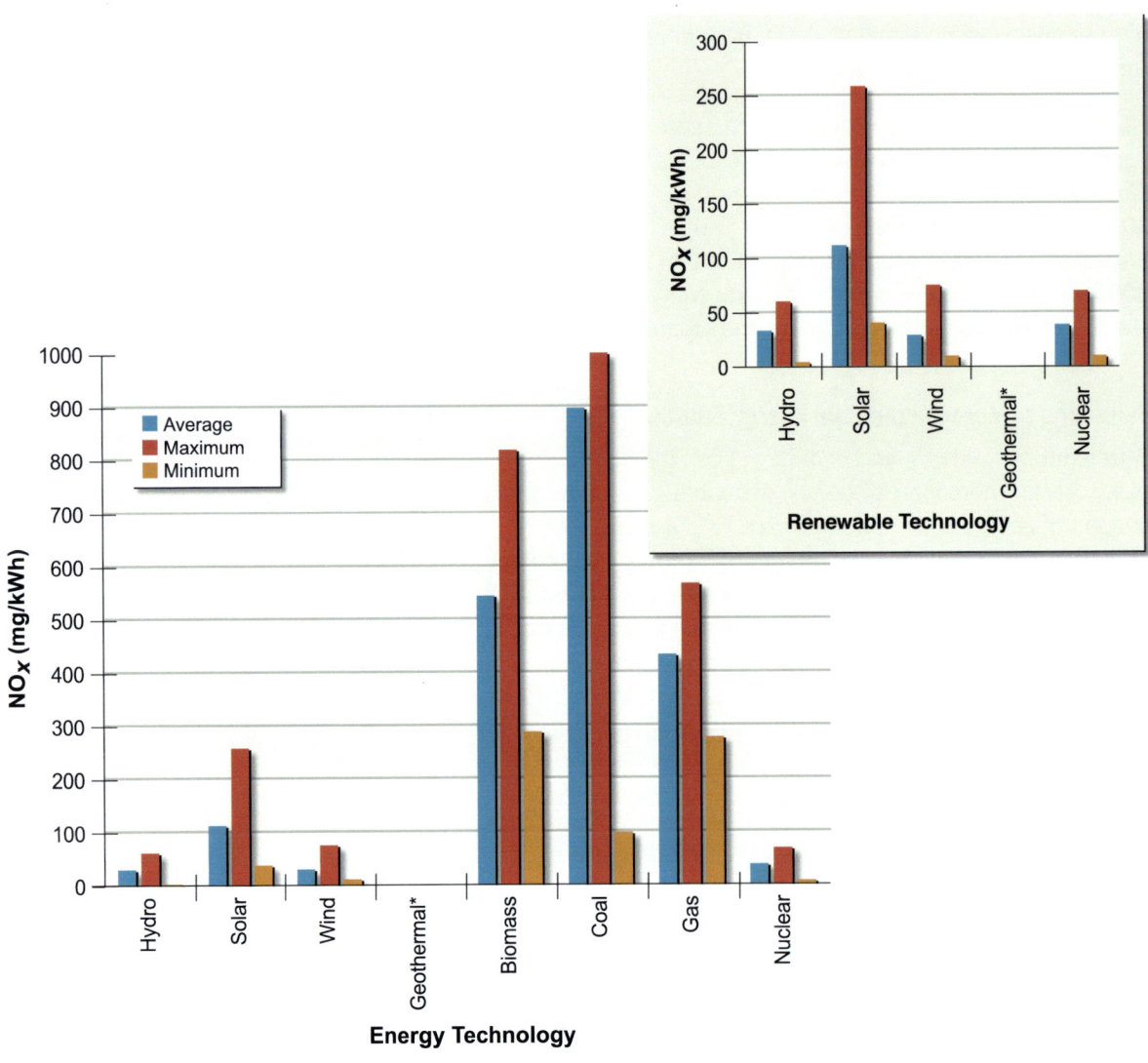

FIGURE 1-8. Life cycle of nitrogen oxide (NOx) emissions (top) and life cycle of land use per kilowatt-hour for various electricity sources (bottom). An important role for environmental engineers will be to compare renewables by conducting life-cycle analyses of all impacts such as land use, water use, and pollution.

Environmental impacts also accrue in the production of renewable energy components, including turbine blades, photovoltaic cells, and electronics, which require energy and materials to produce with associated land, water, and air impacts. Production of some components, such as photovoltaic cells, generate toxic substances that may contaminate land or water resources.[79] Wind turbines may use rare earth minerals, most of which have been mined using processes that result in substantial environmental pollution.[80]

The use of biofuels such as corn-based ethanol for transportation has implications across food, water, and energy systems. Biofuels derived directly from plants, have implications for land and water use and crop prices.[81] Biofuels can also be harvested from algae or produced indirectly from agricultural, commercial, domestic, and/or industrial wastes (see also Challenge 3). In China, over 40 million household-scale anaerobic digesters have been installed that use bacteria to convert plant and animal waste to methane gas.[82] Using anaerobic digestion, environmental engineers have an opportunity to design and create distributed energy systems that also reduce pollution. Analysis of environmental impacts and benefits across the full life cycle of renewable technologies, including the energy return on investment, is a growing role and opportunity for environmental engineers.

Finding Ways to Get Energy Where It Is Needed

Providing energy to the one in seven people who do not yet have it will require decentralized solutions. With declining costs, renewable energy technologies are offering cost-effective alternatives in remote locations compared to centralized systems, replacing traditional energy sources that generate harmful air pollutants, such as diesel generators and biomass burning.[83] Continued advances in transmission and storage as well as further reductions in cost will help provide access to reliable renewable energy supplies.

The use of renewable "microgrids" has emerged as a promising solution to sustainably supply locally-generated electricity to remote regions that are not connected to a conventional power grid. Microgrids can use solar panels, wind, or hydropower to provide cleaner, more cost-effective electricity at a community scale, with generators and battery technology providing backup power when needed. Alaska has been a leader in the development of microgrids, building them on top of the many diesel generators that have served the state's remote areas since the 1960s. Today, Alaska's 70 microgrids comprise about 12 percent of renewably powered microgrids in the world.[84] In urban areas, microgrids can provide backup power during natural disasters, such as Hurricane Sandy, which damaged large parts of the power grid in the Northeast.[85] Projects to build microgrids are accelerating across Asia, Latin America, and Africa. For example, the University of Chile is working to extend the 10-hour capacity of a small diesel-powered electrical grid in the Andes Mountains by supplementing it with solar photovoltaic, wind energy, and a battery system.[86]

In low-income countries, the use of microgrids and smaller, stand-alone systems (such as solar home systems) presents a significant opportunity for providing energy to rural populations without centralized power supply. To achieve universal access to energy by 2030, the International Energy Agency estimates that an additional 340 million people in low-income countries would need to be connected to microgrids, with another 110 million using stand-alone energy systems.[87]

Middle- and high-income countries are challenged to incorporate renewables into the operations of the traditional electrical grid, and significant modification of the grid will be needed.[88] Renewable energy is not necessarily generated where it is needed, and unlike fossil fuels, sunshine, wind, and geothermal energy cannot be transported. Therefore, large-scale transmission projects may also be required. For example, most wind power in the United States is generated in low-population High Plains states, which has prompted proposals for large-scale transmission projects to bring this electricity to population centers in the Midwest and eastern parts of the country.

Energy storage is another challenge, given that solar- and wind-driven electricity production is intermittent. When there is too little sun or wind, production can fall short of demand, while an abundance of sun and wind can create too much electricity that has to be used or curtailed to avoid overloading the grid. Ideas being discussed include creating a bigger grid, or "supergrid," to increase the probability that the sun will be shining or the wind will be blowing in one part of a supply network, if not another.

Many efforts are focused on the development of cost-effective energy storage technologies to smooth out the intermittent nature of solar and wind energy, enabling renewables to provide a much larger percentage of the energy portfolio. Innovation in this realm includes the use of large hydroelectric dams to store electric energy from wind and solar installations in the form of potential energy (see Sidebar). A similar idea is to use electricity during periods of low demand to

USING THE HOOVER DAM FOR ENERGY STORAGE

The growth of solar and wind power is fueling new ideas about how to store excess electrical energy for use when there is not enough solar- and wind-driven energy to meet demand. In 2017, the Los Angeles Department of Water and Power proposed using the Hoover Dam for energy storage to provide greater flexibility and reliability to an electrical grid that is increasingly reliant on renewable energy.[91] Built in the 1930s for flood control, irrigation, and hydroelectric power, the dam sends water stored in Lake Mead through turbines to provide electricity to about 1.3 million people in California, Nevada, and Arizona. From there, the water flows down the Colorado River where it is no longer available to the hydropower plant for making electricity.

The proposed plan is to build a pump station about 20 miles downstream of the dam. Powered by surplus electricity generated by solar and wind energy, the pump would capture river water from the lower Colorado and send it back up to Lake Mead where it can be used to generate electricity when demand exceeds supply. In essence, the process would allow the dam to store solar- and wind-derived electrical energy in the form of potential energy, acting like a giant storage battery.

In general, the relative economic advantage offered by pumped storage at hydroelectric dams makes it the most widely used method for the large-scale storage of electrical energy.[92] However, it is important to weigh all of the benefits and costs when applying that technology to particular hydroelectric dams. Consideration of costs includes the potential ecological impacts associated with water fluctuations in rivers related to the energy storage efforts. What effects would those fluctuations have on the diversity and ecological function of plants and animals in and near the river? In addition, there are potential recreational and aesthetic impacts to humans in the proximity of the pumped storage system. Environmental engineering expertise will be needed to consider the full life-cycle impacts of alternative energy storage solutions.

pump ambient air into a storage container and, when electricity is needed, allow the compressed air to expand to drive turbines.[89] Other promising leads in this vein include mechanical storage with rail or flywheels, and use of excess electricity to create other fuels, such as hydrogen.[90]

What Environmental Engineers Can Do

Environmental engineers bring decades of experience in water treatment and alternative water supply technologies to address challenges ahead related to water scarcity. Environmental engineers have traditionally had less experience in issues of food supply and energy. Nevertheless, opportunities abound for environmental engineers to apply systems thinking (see Box 1-1) to analyze the interrelated behaviors of water, food, and energy systems and their interaction with the environment that supports them. Through systems and life-cycle thinking, engineers can help develop technologies and strategies to sustainably supply food, water, and energy to Earth's growing population (see Box 1-2 for examples).

Addressing this challenge will require convergence of multiple disciplines across behavioral and social sciences, engineering, and science. Environmental engineers can work in collaboration with experts in agriculture, energy, health, ecology, molecular biology, data science, social science, policy, and other disciplines.

BOX 1-1. SYSTEMS THINKING

We now face environmental issues that are global, complex, and interconnected. Environmental engineers are trained to bring a systems-based view to problem solving, allowing for more innovative and appropriate solutions. For example, environmental engineers understand the movement of contaminants between air, water, and land so that they can develop methods to reduce pollution in one sector that do not result in adverse consequences in another. Environmental engineers consider a broad array of issues that often involve systems of systems, such as the vital role and value of ecological services as well as the life cycle impacts and benefits of an engineered system, from its raw materials to end of life.[93]

Although environmental engineers have a long history of thinking about complex environmental systems, there is a need to routinely extend this type of thinking beyond the natural world to encompass broader aspects, such as the regulatory environment, economic drivers, and social behavior. For example, through systems thinking, environmental engineers can also consider the specific needs and perspectives of disadvantaged groups and understand the role of economic incentives and policy instruments to align socioeconomic behavior with environmental goals.[94]

Environmental engineers work on systems that are integrated and complex, including technical aspects as well as social, environmental, and economic facets. These complex systems are difficult to predict in that they are nonlinear, have feedback mechanisms, are adaptive, and have emergent behavior.[95] Only recently has computing power increased sufficiently to enable quantitative evaluations of technological advances in the context of potential changes in underlying social and economic systems.[96] With these tools, environmental engineers can help design solutions that are appropriate, effective, and sustainable.

BOX 1-2. EXAMPLE ROLES FOR ENVIRONMENTAL ENGINEERS TO HELP SUPPLY FOOD, WATER, AND ENERGY FOR EARTH'S GROWING POPULATION

Environmental engineers have many strengths to help address the challenge of supplying clean water and nutritious food to Earth's population in the 21st century. Examples include

Food
- Develop a systems-level "farm to plate" assessment to identify ways to reduce waste, energy, and water consumption and to improve access to healthy food choices.
- Develop precision delivery systems for water, nutrients, and pesticides to minimize impacts on air quality, soil, groundwater, and ecosystems while reducing waste and energy consumption.
- Develop on-site systems to affordably transform agricultural waste into energy.
- Assess the costs and benefits of alternative food sources, such as cultured meat, from human and environmental perspectives.
- Develop aquaculture and aquaponics systems to meet increasing demand for seafood to reduce impacts on ocean supplies with integrated nutrient recovery and reuse to minimize adverse effects on the environment.
- Design urban agriculture systems to utilize waste energy and recycle water, minimizing water use and pollution.

Water and Sanitation
- Considering the full spectrum of human development conditions, develop energy-efficient water conservation strategies and technologies that are socially acceptable and implementable.
- Develop low-cost desalination and water reuse technologies, including strategies to reduce energy use and manage or reuse waste streams to minimize environmental impacts.
- Develop water supply and water quality forecasting tools, including low-cost, distributed sensing systems, to anticipate water availability and quality threats.
- Develop and evaluate energy-neutral or energy-positive cost-effective wastewater treatment technologies suitable for low-, middle-, and high-income settings that provide enhanced contaminant removal, minimize energy consumption, and promote safe water reuse.
- Participate in innovative interdisciplinary teams to develop and evaluate approaches to water, sanitation, and hygiene challenges in low-income countries.
- Develop improved diagnostic tools and predictive modeling approaches to understand the state of aging water infrastructure and develop cost-effective strategies to maintain the water services provided by existing infrastructure.

Energy
- Conduct life-cycle analyses of renewable technologies and distribution strategies in terms of benefits provided and water and energy use and pollution, including all stages. Develop approaches to minimize those impacts.
- Investigate approaches to store energy, such as with hydroelectric dams or batteries, and examine associated environmental impacts and ways to minimize those impacts.
- Develop low-cost ways to reduce environmental impacts associated with traditional energy production.
- Develop viable, sustainable biofuel options.

GRAND CHALLENGE 2:

Curb Climate Change and Adapt to Its Impacts

It is now more certain than ever that humans are changing Earth's climate.[97] The burning of fossil fuels for electricity generation, transportation, heating, cooling, and other energy uses has raised the concentration of global atmospheric carbon dioxide (CO_2) to more than 400 parts per million (ppm)—a level that last occurred about 3 million years ago when both global average temperature and sea level were significantly higher than today.[98] At the same time, the production of fossil fuels and agricultural and industrial processes also have emitted large amounts of methane and nitrous oxide, both powerful greenhouse gases, into the atmosphere.

The heat trapped by the sharp rise in greenhouse gases has increased Earth's global average surface temperature by about 1.8°F (1.0°C) over the past 115 years, and at an increased rate since the mid-1970s (see Figure 2-1).[99] This warming has been accompanied by rising sea levels, shrinking Arctic sea ice, reduced snow pack, and other climatic changes. Many urban areas across the globe have witnessed a significant increase in the number of heat waves. More rain is falling during the heaviest rainfall events, causing flooding and further stressing low-lying coastal

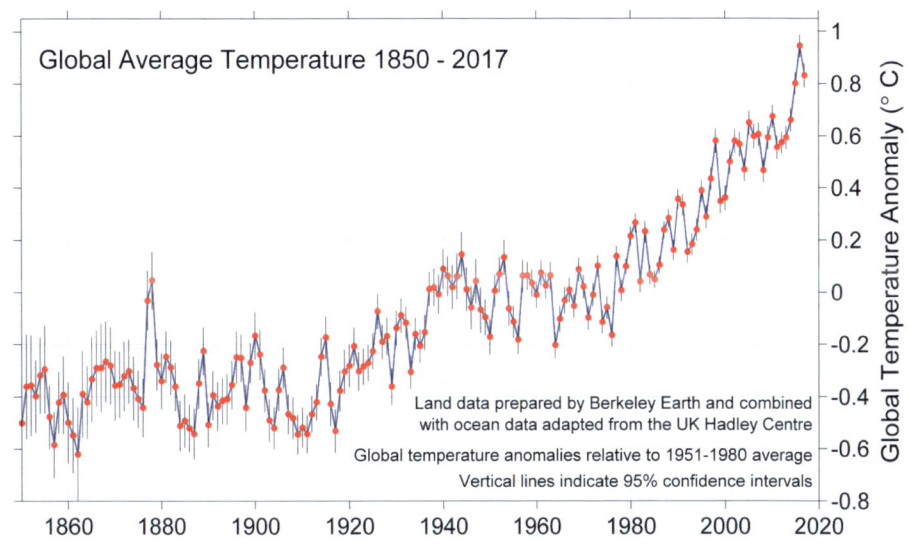

FIGURE 2-1. Earth's global average surface temperature has risen about 1.8°F (1.0°C) over the past 115 years, with much of that increase occurring since the mid-1970s. The temperature changes (anomalies) are relative to the global average surface temperature of 1951–1980.

zones already vulnerable to storm surges and other causes of temporary coastal flooding, along with sea-level rise.[100] In other areas, prolonged dry periods and droughts are increasing the risk of destructive wildfires and water shortages.

If greenhouse gas emissions continue to rise in the 21st century, Earth is expected to warm by an additional 4.7°F to 8.6°F (2.6°C to 4.8°C) by 2100 (relative to 1986-2005).[101] The greater the warming, the greater the impacts will be. In the United States, each degree of warming (Celsius) is projected to result in a 3 to 10 percent increase in the amount of rainfall during the heaviest rain events, a 5 to 15 percent reduction in the yields of crops as currently grown, and a 200 to 400 percent increase in the area burned by wildfire in western states.[102] Similar types of changes are expected in many other parts of the world, which could be most devastating to low-income countries that do not have the resources to respond or adapt.[103]

Warming of about 5.4°F (3°C) or more could push Earth past several "tipping points." For example, this amount of warming could melt the Greenland ice sheet, which would raise global average sea level an additional 20 feet (6 meters).[104] It could also accelerate the thawing of permafrost, which would accelerate the release of CO_2 and methane stored in frozen soil, exacerbating warming.[105] While projections such as these are useful in planning for the changes ahead, it is also important to recognize that a great deal remains unknown, particularly when it comes to the complex feedbacks among human activities, ecosystems, and the atmosphere.

For decades, scientists have led the efforts to understand and predict climate change effects, but engineers are now recognizing that their efforts are needed to help develop and implement solutions. Conceptually, climate solutions are divided into two areas of focus: mitigation and adaptation. Mitigation refers to efforts to reduce the magnitude or rate of climate change by reducing emissions of carbon dioxide and other greenhouse gases or removing them from the atmosphere. Adaptation refers to solutions that avoid or lessen the impacts of climate change on people, ecosystems, resources, and infrastructure. Environmental engineers have an opportunity to be leaders in developing technologies and systems that provide solutions on both of these fronts. Given that future climate changes likely hold surprises, it will be important to remain nimble, incorporate new knowledge, and work to address uncertainty as environmental engineers develop, test, and implement solutions.

Reducing the Rate and Magnitude of Climate Change

A sharp reduction in emissions of greenhouse gases to the atmosphere is needed to slow climate change and prevent some of the most severe impacts. For the past few decades, international climate talks have focused on establishing goals to minimize the planet's warming, with the most recent goals set at limiting future warming to 3.6°F (2°C) above preindustrial levels. The 2016 Paris Agreement set an aspirational target of limiting warming to 2.7°F (1.5°C). Since the planet already has warmed about 1.8°F (1°C), scientists have calculated that, in order to stay within the 3.6°F (2°C) limit, atmospheric CO_2 concentrations must not rise beyond 450 ppm, which in turn requires 40 to 70 percent reductions in global anthropogenic greenhouse gas emission by 2050 compared to 2010, and emissions levels near zero or below in 2100.[106]

A special report issued in 2018 by the Intergovernmental Panel on Climate Change urged world leaders to work toward limiting warming to 2.7°F (1.5°C) to avoid the severe impacts on weather extremes, ecosystems, human health, and infrastructure that are expected to occur at 3.6°F (2°C) warming.[107] Meeting that tougher goal will require global emissions to be reduced by about 45 percent from 2010 levels by 2030, reaching net zero emissions by 2050. Meeting those emissions targets will require dramatic reductions in global CO_2 emissions combined with the active removal of CO_2.[108]

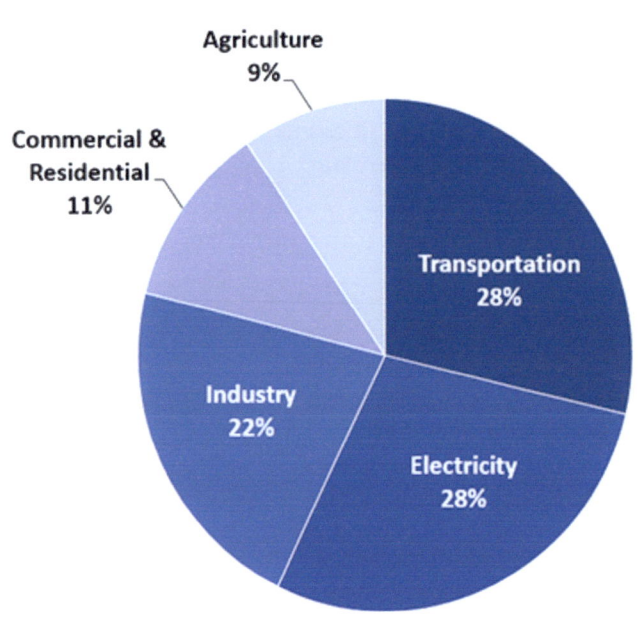

FIGURE 2-2. Total U.S. greenhouse gas emissions by sector in 2016, in CO_2 equivalents.

Greenhouse gas emissions are driven by the use of energy for electricity generation, transportation, industry uses, commercial and residential needs, and agriculture. Figure 2-2 shows the U.S. breakdown for greenhouse gas emissions sources. Emissions can be reduced by using energy more efficiently, switching to fuels that produce less (or no) greenhouse gases, and capturing the emissions before they enter the atmosphere.

In general, reducing emissions will require that existing and planned transportation, building, and industrial infrastructure be converted to electricity that is generated with substantially lower carbon intensity. Doing so will have the added co-benefit of reducing the environmental and human health impacts associated with coal, oil, and natural gas extraction and fossil-fuel-generated electricity (see Challenge 3).[109]

Using Energy More Efficiently

High-income countries have already substantially reduced their energy use per capita and per unit of economic output. These improvements have resulted from significant technological changes, such as the advent of LED lighting, energy-efficient appliances, and other efficiency intelligence in buildings; industrial restructuring to enhance productivity; and investment in fuel-efficient transportation technologies. Lower- and middle-income countries are beginning to make similar gains.

Efficiency gains made to date, however, will not be sufficient to avoid a 3.6°F (2°C) average rise in global temperatures. More than 80 percent of vehicle miles traveled in 2050 need to be powered by something other than an internal combustion engine.[110] Substantial efficiencies also are needed in industry and in the heating and cooling of buildings. In Germany, for example, a high-level commission calculated that German buildings would need to achieve a 54 percent improvement in efficiency by 2030 to meet stated emission reduction goals.[111] Effectively deploying new and emerging technologies can help advance these goals. It has been estimated that energy-efficient technologies for residential and commercial buildings, transportation, and industry that exist today or are expected to be developed soon could reduce U.S. energy use by 30 percent, slashing greenhouse gas emissions along with other air pollutants, while also saving money.[112]

Switching to Fuels That Produce Less (or No) CO_2

As discussed in Challenge 1, there are many sources of energy that produce little or no CO_2 emissions, including solar, wind, geothermal, and hydropower. Although low-emissions energy sources exist, there is still a long way to go toward their widespread adoption. As of 2017, U.S. electricity generation was composed of about 63 percent fossil fuels, 20 percent nuclear, and 17 percent hydropower and other renewables.[113] A study by the Department of Energy's National Renewable Energy Laboratory shows that it is feasible for the United States to generate most of its electricity from renewable energy by 2050, but a number of challenges remain.[114] Cost has been a significant barrier, although costs are dropping for both solar and wind power technologies.[115]

Reducing U.S. emissions enough to stay within the 2.7°F (1.5°C) limit would require the current balance of energy production to shift substantially, such that 70-85 percent of electricity is generated from noncarbon-emitting sources.[116] In China, maturation and economic restructuring of the industrial sector has already substantially reduced coal consumption per unit of output, a trend that is projected to continue and be further enhanced by their recently introduced carbon cap and trade system.[117] In addition, China is leading the charge in developing renewable energy, for example, building 45 percent of the world's solar installations in 2016.[118]

Advances are needed to improve the efficiency and reduce the costs of such energy sources to make them competitive with traditional fossil fuel–based sources. In addition, since many renewables produce energy intermittently, there is a need for energy storage systems with increased capacity, scalability, reliability, and affordability, as discussed in Challenge 1.

Nuclear power is one low-emission energy source that already comprises one-fifth of U.S. electricity generation. Increasing the use of nuclear power could help reduce carbon-emitting energy generation, but there are significant barriers, including cost, public concerns related to safety and waste disposal, the high business and regulatory risks involved in designing and building nuclear power plants, and the lack of progress in developing long-term waste repositories. Retiring existing nuclear plants will exacerbate the challenge of reducing CO_2 emissions from the power system, because large increases in renewable and other zero-emitting energy sources will be needed simply to replace zero-emitting nuclear energy. To support continued nuclear capacity, working in combination with renewables, research is needed on advanced nuclear technologies for next generation reactors designed to significantly improve performance and safety.[119]

Moving to electrically powered transportation with increased renewable energy generation would substantially reduce fossil fuel use, because more than 90 percent of the transportation fuels are petroleum based.[120] Electric vehicle technology has advanced substantially in the past 5 years, with roughly 2 million all-electric and plug-in hybrid vehicles on the road worldwide today,[121] and automobile companies are increasing investments in electric vehicle production. For example, Volvo announced a plan to transition all of the company's car models to electric or hybrids by 2030, Ford has announced an $11 billion investment in electric vehicles, and GM plans to release 20 new models of electric vehicles by 2023.[122] Several countries, including Britain, France, and Norway, cities such as Beijing, and several U.S. states have proposed banning gasoline- and diesel-powered cars as early as 2030.[123] Achieving the transition to electric-based transportation systems raises many engineering challenges beyond the need for low-carbon energy sources, including the need for charging infrastructure, better battery performance, and faster recharge times.

Making progress toward reducing emissions will depend in large part on private-sector investments and on the behavioral and consumer choices of individual households, which are explored in more detail in Challenge 5. Governments at federal, state, and local levels can influence those choices through policies and incentives. Such policies can include setting a price on emissions, such as a carbon tax or cap-and-trade system; providing information and education on voluntary emission reductions; and mandates or regulations designed to control emissions, for example, the Clean Air Act, automobile fuel economy standards, appliance efficiency standards, building codes, and requirements for renewable or low-carbon energy sources in electricity generation.

Advancing Climate Intervention Strategies

Even if human-caused carbon dioxide emissions were to cease today, it would take millennia for natural processes to return Earth's atmosphere to preindustrial carbon dioxide concentrations.[124] To avoid the worst impacts of warming, it is no longer enough to reduce emissions. Deploying negative-emission technologies that remove carbon dioxide from the atmosphere and reliably sequester it will also be needed.[125]

Some carbon dioxide removal strategies focus on accelerating natural processes that take up carbon dioxide. Changes in agricultural practices can enhance soil carbon storage, for example, by planting fields year-round in crops or other cover crops.[126] Land use and management practices can be employed that increase the amount of carbon stored in terrestrial environments, such as forests and grasslands and in near-shore ecosystems, such as mangroves, tidal marshes, and seagrass beds.[127] One recent study estimates that nature-based approaches can deliver more than one-third of the carbon reductions needed by 2030 to stay within the 3.6°F

(2°C) limit at competitive costs,[128] but there are many unknowns. Further research is needed to determine what conditions and practices can maximize carbon uptake in plants over the long term. There can also be unintended effects. For example, planting more trees in northern boreal forests can contribute to warming, because in winter months the trees can obscure snow that reflects sunlight.

Other technologies being explored seek to actively remove CO_2 from the atmosphere and from point sources and sequester it. One technology involves growing plants such as switchgrass to be converted to fuel, coupled with capturing and storing any CO_2 emissions from biofuel burning (called bioenergy with carbon capture and sequestration, or BECCS). Another approach proposes using chemical processes to capture CO_2 directly from the air and concentrate it for storage (called direct air capture and sequestration, or DACS). These technologies will be needed around the world because many countries will still be using significant amounts of fossil-fuel-generated electricity by 2050. They will also be needed to mitigate emissions where electrification is not possible and for industrial processes that produce carbon dioxide.

Engineering challenges in carbon removal strategies include the need to reduce costs, increase the scale of the technologies, and store or reuse the carbon in ways that keep it from being released back into the atmosphere. Available land is a key limiting factor for the potential of removing CO_2 through reforestation or growing fuel crops; removing 10 gigatons CO_2 per year (about one quarter of global yearly emissions) by 2050 would require the use of hundreds of millions of hectares of arable land.[129] Land use at that scale could threaten food security, given that food demands are expected to increase by 25 to 70 percent over the same time period.[130] Breakthroughs in agriculture discussed in Challenge 1, including advances in crop productivity, alternative methods of growing food, food waste reduction, and changes in diet, will be needed.

A different set of climate intervention strategies seeks to reduce warming by reflecting sunlight off of specially treated clouds and aerosols. In general, such technologies are not as developed as carbon dioxide removal strategies and carry greater risks of unintended consequences that are not well understood.[131]

Reducing Other Greenhouse Gases

Methane, nitrous oxide, and some industrial gases (e.g., hydrofluorocarbons) comprise about 18 percent of U.S. greenhouse gas emissions in terms of CO_2 equivalents.[132] Molecule for molecule, those gases are much stronger climate warming agents than CO_2, although they are less abundant, and some do not last as long in the atmosphere. Methane, for example, is about 28 times more potent as a greenhouse gas compared to CO_2, making it particularly important to prevent or capture methane leaks from oil and gas systems, coal mines, shale gas extraction, and landfills.[133] To that end, there is a need for better systems and methodologies to measure and track methane leakage throughout those systems.[134]

Agriculture is one of the largest sources of non-CO_2 greenhouse gases. Methane is produced when livestock digest their food and also is emitted in large quantities

from rice paddies. Nitrous oxide arises from the use of nitrogen fertilizers. Precision agriculture techniques can help farmers minimize fertilizer use and reduce nitrous oxide emissions (see also Challenge 1). Feeding livestock easier-to-digest foods and strategically managing livestock waste—through proper storage, reuse as fertilizer, and recovery of methane—also can help reduce emissions.[135] Efforts to curb agricultural methane emissions can benefit from new insights and biotechnology tools that offer new ways to study the complex microbial ecosystems involved in soils, manure management, and livestock digestion.

Some short-lived pollutants that are not greenhouse gases also contribute to warming. One example is black carbon, commonly called soot, which absorbs sunlight and traps heat in the atmosphere. Black carbon is produced by incomplete fuel combustion and burning of biomass (e.g., the dung used in cookstoves). Black carbon also can amplify regional warming by leaving a heat-absorbing black coating on otherwise reflective surfaces, such as snow in mountainous regions. Although North America and western Europe were the major sources of soot emissions until about the 1950s, low-income nations in the tropics and East Asia are the major source regions today. Identifying and targeting the largest sources of black carbon could be crucial to curbing warming in the short term.

What Can Environmental Engineers Do to Curb Climate Change?

Environmental engineers have an opportunity to be leaders in developing technologies that will help slow warming through alternative energy development, green infrastructure, carbon capture and sequestration, and monitoring and measurement, as summarized in Box 2-1. Although the challenge to curb climate change will stretch environmental engineering beyond its traditional boundaries, many of the skills typical of environmental engineers can be applicable for advancing these goals. For example, the design of technologies to capture and store carbon underground, in soils, and in coastal ecosystems can take advantage of environmental engineers' expertise in water chemistry, environmental microbiology, groundwater and surface water hydrology, and atmospheric chemistry. Environmental engineers can also bring large-scale perspectives to illuminate how proposed technologies will interact with multiple systems. Specific applications of those skills might include

- Using the tools of geochemistry to engineer accelerated mineralization processes that would transform carbon into a stable carbonate, while avoiding water quality impacts.
- Using the emerging tools of synthetic biology and microbial ecology to abate greenhouse gas emissions and generate chemicals, materials, and fuels.
- Using the tools of life-cycle assessment to explore efficiencies for producing low-carbon liquid fuels from biomass feedstocks without increasing overall water use.
- Using the tools of life-cycle assessment to assess and optimize the energy return on investment (the ratio of the amount of usable energy delivered from a particular energy resource to the amount of energy used to obtain that energy resource).

BOX 2-1. EXAMPLE ROLES FOR ENVIRONMENTAL ENGINEERS TO HELP CURB CLIMATE CHANGE

Environmental engineers can play an important role in collaboration with other disciplines to address four areas related to slowing climate change.

Increasing Energy Efficiency
- Using life-cycle analysis, identify opportunities for improved energy efficiency across sectors to focus energy efficiency improvements toward those with the greatest benefits.
- Identify opportunities for the use of the heat that is a by-product of the generation of electricity. Currently much of this heat is "wasted" during cooling processes.

Advancing Alternative Energy Sources
- Identify opportunities for addressing environmental issues associated with promising renewable energy sources, including hydropower, solar, and wind.
- Develop low-cost reliable anaerobic carbon conversion systems to turn organic wastes, including human waste as well as agricultural plant and forest residues, into energy.
- Develop strategies to manage nuclear waste.

Advancing Climate Intervention Strategies
- Develop biological and mechanical carbon capture methods that can be scaled at reasonable cost.
- Develop uses for captured carbon and methods for safe storage, including monitoring for leakage.
- Improve understanding of the factors that influence the permanence of carbon capture by vegetation and soils.

Reducing Other Greenhouse Gases
- Develop monitoring tools to detect emissions of methane in natural gas systems and methods to minimize or eliminate them.
- Develop technologies and approaches to reduce greenhouse gas emissions from agriculture.
- Identify the largest sources of black carbon and develop low-cost strategies to reduce these emissions.

Adapting to Climate Change Impacts

Many of the things people use or do every day—from roads to farms, buildings to subways, jobs to recreational activities—were optimized for the climate of the 19th and 20th centuries. They were built with the assumption of certain temperature ranges, precipitation patterns, frequency of extreme events, and other manifestations of climate, which are now shifting. Even if humankind were to succeed in limiting global climate change in accordance with current goals, adaptation will be needed to protect people, ecosystems, infrastructure, and cultural resources from the impacts of climate change, many of which are already evident.

Sea level is one area in which those impacts are already being felt. Since 1900, global mean sea level has risen about 8 inches, driven by expansion of the warming ocean, melting of mountain glaciers, and losses from the Greenland and Antarctic ice sheets.[136] This rise has caused coastal cities to see an uptick in flooding, both during storms and as "sunny-day" flooding from tides alone. These flooding events disrupt economies, make it difficult to deliver emergency services, and disproportionately affect older, infirm, and low socioeconomic status populations.

Global sea level is expected to rise by an additional 0.5 to 1.2 feet by 2050 and 1 to 4.3 feet by 2100, which will increase the frequency and severity of flooding (see Figure 2-3). Even at the low end of that estimate, up to 200 million people could be affected worldwide and 4 million people could be permanently displaced as frequent or permanent flooding makes low-lying developed areas uninhabitable.[137] Some communities already are being forced to relocate as a result of sea-level rise, including Native American communities in Alaska, communities south of New Orleans in the Louisiana Delta and island communities in the Pacific and Indian oceans. In addition to flooding, sea-level rise causes erosion and saltwater encroachment, which kills forests near the coasts, reshapes marshes and wetlands, and renders aquifers along the coast unusable for human consumption without desalination technology.

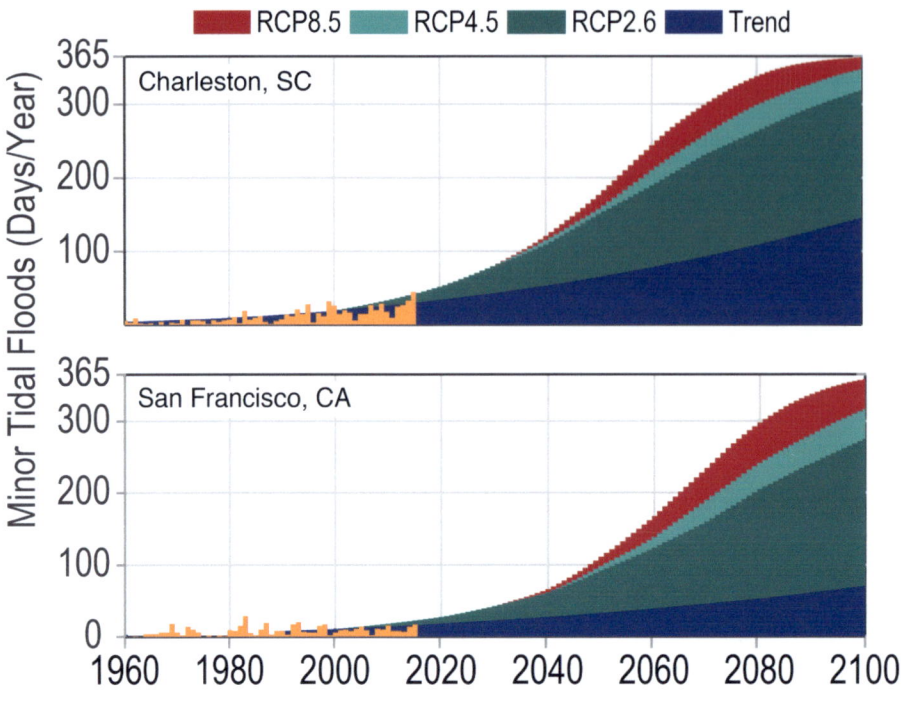

FIGURE 2-3. Annual occurrences of tidal flooding, also called sunny-day or nuisance flooding. Recent documented events are shown in orange and future flooding projections based on three greenhouse gas emission scenarios known as representative concentration pathways (RCP) ranging from low (RCP2.6) to high (RCP8.5).

Climate change is also expected to intensify regional contrasts in precipitation that already exist: Dry areas are expected to get drier and wet areas to become even wetter. Changes in precipitation patterns have resulted in heavier rainfalls, reduced snow cover, and glacial extent, and doubled the amount of land area classified as "very dry."[138] Warmer temperatures tend to increase evaporation from oceans, lakes, plants, and soil, exacerbating the impacts in areas of reduced precipitation.

Extreme precipitation events are becoming more frequent, leading to increased flooding as well as spikes in the release of some pollutants during heavy storms.[139] In August 2016, for example, more than 2 feet of rain fell in central Louisiana over 10 days, an event the National Weather Service called a "one in a thousand year" event. Scientists predict that climate change will cause an increase in the number of the most severe hurricanes, leading to stronger storm surges and more intense rainfall events.[140] In 2017, Hurricane Harvey dumped a staggering 50 inches of rain on Houston, which is as much rain as typically falls there over an entire year. Work is ongoing to assess the future probability of similar rainfall events.

As Earth's climate warms, changing temperatures are expected to reduce agricultural productivity for some major crops and may exacerbate the impacts of agricultural pests and pathogens.[141] Extreme heat waves will become more frequent, causing additional wildfires and further degrading air quality. Urban residents, especially those without access to air conditioning, are vulnerable to heat waves, as heat island effects make building and pavement surfaces 7°F to 22°F (4°C to 12°C) warmer than the surrounding natural environment.[142]

These changes are expected to pose a number of serious risks to human societies, affecting freshwater management, ecosystems, biodiversity, agriculture, urban infrastructure, and human health. To manage the risks and lessen the impacts, there is an urgent need to develop and deploy adaptation measures. Appropriate adaptation measures will vary from location to location, and some climate change impacts will be beyond the scope of adaptation. In some places, incremental steps will be sufficient to manage risk over the next several decades. In other places, transformative changes, such as relocation, are likely to be required. Because there is a great deal of uncertainty regarding future changes, advances in tools that support robust decision making under deep uncertainty[143] and adaptive management—a model that maximizes flexibility as new knowledge becomes available—will be crucial.

Adaptation strategies range from technological and engineered solutions to social, economic, and institutional approaches. Social and cultural factors will affect which strategies are acceptable to local communities. The following examples highlight current strategies being developed and future areas of focus for adaptation. Other examples related to water scarcity are discussed in the context of Challenge 1.

Building Disaster Resilience

Communities need to increase their resilience to disasters, such as floods and wildfires, which are expected to become more frequent and more intense in the decades ahead. Flood impacts can be lowered by, for example, developing building standards based on future flood risks and curtailing development in high-risk areas. Improved local projections of flood risk based on changes in climate and land use are needed to inform such planning and decision making; advanced GIS technologies are offering flexible tools that engineers and communities can use toward this goal. In a departure from past strategies, which emphasized centralized flood control management with levees and dams that have severe impacts on river and floodplain ecosystems, communities are increasingly turning to natural systems to manage flood risks while enhancing habitat, water quality, and other environmental services.

Resilience is the ability to prepare and plan for, absorb, recover from, and more successfully adapt to adverse events.[146]

Wildfires play a natural role in preserving the health of forests and other ecosystems that are adapted to wildfire. However, growth of communities into the wildland-urban interface and also climate change, which has made fire seasons longer and droughts worse, has increased the costs and impacts of wildfires.[144] California suffered its worst fire season ever in 2017, which was followed by rainstorms that triggered devastating mudslides. Globally, billions of dollars are spent to remediate impacts on human health, property damage, loss of tourism, and the restoration of crucial ecosystem goods and services.[145]

A major need related to wildfire is the creation of improved models and measurements to predict wildfire spread and the transport of wildfire smoke emissions. Other efforts to increase resilience to wildfire include improved landscape design principles and adaptive management to protect assets through tree cultivation, prescribed burning, grazing, and education programs to reduce accidental ignitions.

Reducing Impacts on Ecological Systems and Services

For many aquatic and terrestrial species, climate change has altered habitat conditions, leading to changes in biodiversity and species abundance and distribution. Increasing ocean temperatures and nutrient inputs from rivers are expanding the number and size of areas with low-oxygen conditions ("dead zones"), impacting commercial fisheries. Declining Arctic sea ice is reducing the habitat and hunting ground for polar bears, threatening survival of the species. Some changes are happening too quickly to allow for adaptation. However, efforts to reduce other environmental stressors, such as pollution (see Challenge 3), could reduce the severity of climate impacts and prevent species extinctions. Other adaptation strategies include habitat restoration, assisted migration, active management of invasive species, and updated management strategies for fisheries.[147]

Adapting Agricultural Practices

Technological advances during the 20th century's green revolution dramatically improved agricultural yields, economic stability, and food security in many parts of the world.[148] However, climate change threatens to undercut some of these advances. Agricultural adaptations such as adjusting planting dates, seed or crop selection (for example, to develop more flood- and drought-tolerant crops), or altering irrigation practices have the potential to buffer the impacts of climate change.[149] In the long run, it may be necessary to shift the location of agricultural operations or even to shift human diets (see Challenge 1). Additional economic and institutional strategies will be necessary to maintain food security amid increased weather variability and climate extremes.[150]

Adapting Infrastructure for Sea-Level Rise

Widespread adaptations in infrastructure are needed to adjust to climate change. Adaptation strategies include ensuring that critical infrastructure and systems such as water supply, wastewater, and solid waste management systems, electricity-generating facilities, hospitals, and transportation systems are resilient to expected heat, storm, and flooding stressors. With projections of 1 to 4 feet of sea-level rise by the end of the century,[151] engineers are developing ways to hold back the sea where possible or to buy time until more transformative adaptation strategies, including managed retreat, are developed.

In the near term, the city of Miami, Florida, is spending $400 million to raise streets, build sea walls, and construct pumps to reduce frequent flooding.[152] Natural areas, such as coastal wetlands and mangroves, are being protected or restored to maintain natural buffers against storm surge (see Box 2-2). In the Netherlands, engineers have designed long-term strategies to protect heavily developed areas and accommodate increased flooding in less-developed regions. Innovations include smart dikes with embedded sensors that relay real-time status reports to decision makers and ecologically enhanced dikes to provide habitat for marine organisms.[153]

BOX 2-2. REBUILDING WETLANDS IN LOUISIANA

The wetlands of southern Louisiana, the largest in the United States, are disappearing at an alarming rate. More than 1,900 square miles have been lost since 1930 from natural and human causes. Levees and canals have diverted the flow of sediments from the Mississippi River that once sustained the wetlands, while sea-level rise and natural subsidence continue to affect the coastline. Coastal wetlands, including salt marshes and mangroves, provide habitat for local fisheries and are the first line of protection against hurricanes and storm surge. Without action, the state could lose an additional 2,250 square miles of land over the next 50 years. The 2017 Louisiana Coastal Master Plan,[154] approved unanimously by Louisiana's legislature, focuses on restoring the natural flow of sediments to the wetlands, as well as such projects as marsh creation, barrier island restoration, and oyster reef restoration. Wetland loss is problematic in many places, and environmental engineers can contribute to the design of green infrastructure that helps restore lost ecosystems services and retain habitats at risk from sea level rise.

To inform decision making, cities need comprehensive analyses to understand the adaptation options, their potential impacts, and the benefits and costs of local, regional, national, and private-sector infrastructure investments to manage future risks. It will be particularly important to develop economic and institutional strategies to support low-income and vulnerable communities as these adaptation measures are implemented.

Anticipating and Responding to Health Threats

Climate change has a broad range of implications for human health.[155] Changes in temperature are expected to increase heat-related illnesses and deaths (see Figure 2-4), while increases in ozone and wildfires are expected to worsen air pollution, with major effects on human health. Temperature changes may directly affect the transmission of vectorborne and zoonotic diseases carried by rodents and insects, such as ticks and mosquitoes, by increasing the frequency and shifting the geographic areas at risk. Changes in temperature and precipitation patterns may also affect the prevalence or distribution of foodborne, waterborne, and water-related diseases.[156] Temperature changes can also affect wildlife migration patterns, potentially leading to more human-wildlife contact and increasing the risk of infectious diseases that originate in animal populations and spread to humans.

The risk of infectious disease outbreaks also can rise in mass displacement events, such as natural disasters. In the aftermath of Hurricane Maria in 2017, Puerto Rico grappled with many health issues including an outbreak of leptospirosis a bacterial disease.[157] Outbreaks in such settings pose enormous challenges for policy makers and medical, public health, and environmental health personnel, and such events can also contribute to food and water insecurity and malnutrition and cause stress to those who are displaced from their homes.

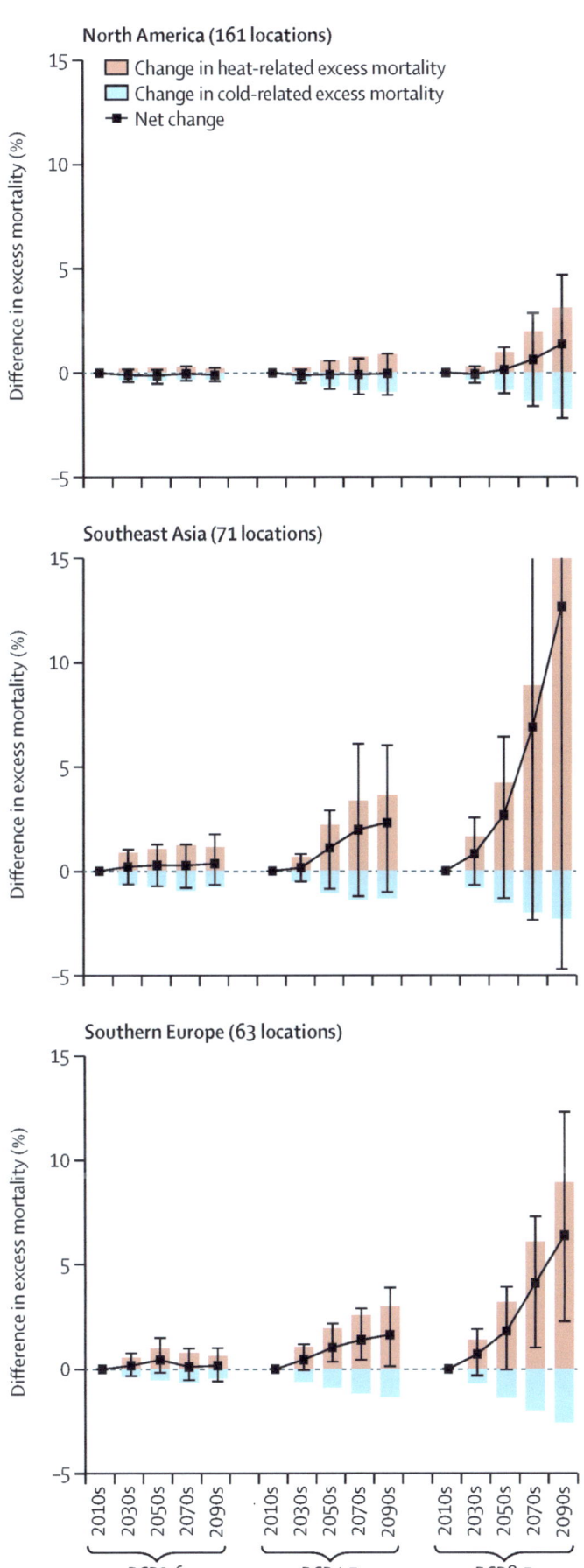

FIGURE 2-4. Projections of temperature-related excess mortality in cities in North America, Southeast Asia, and southern Europe under low-, medium-, and high-greenhouse gas emission scenarios, termed representative concentration pathways (RCP2.6, RCP4.5, and RCP8.5, respectively).

Adaptation strategies could include strengthening infectious disease surveillance systems, developing rapid point-of-care diagnostic tests, and improving rapid response capabilities for disasters and infectious disease outbreaks. Progress toward ensuring water and food security and reducing air and water pollution would also reduce the human health impacts from climate change. One strategy for adaptation in urban areas is to mitigate the urban heat island effect, with efforts needed to test and evaluate the potential for reflective surfaces, vegetation, and other features to reduce the temperature of cities.

What Environmental Engineers Can Do to Advance Climate Change Adaptation

Responding to climate change is about making choices amid substantial uncertainty. Decision strategies have been developed to support robust planning and decision making under deep uncertainty.[158] To support these decision processes, engineers and scientists can improve the understanding of potential long-term climate impacts and examine and communicate the effectiveness and consequences of adaptation strategies, considering a wide array of environmental, social, and economic factors (see also Challenge 5). Environmental engineers are trained with a broad, systems view, which enables them to become a vital bridge across disciplines and act as integrators of information. Using modeling and decision support tools, environmental engineers can work with diverse interdisciplinary teams to synthesize information, analyze adaptation alternatives, and weigh the costs, benefits, and risks. Environmental engineers have skills in uncertainty analysis and can support iterative risk management approaches to analyze climate adaptation strategies for effectiveness and lessons learned in the context of an evolving understanding of climate science. Examples of specific opportunities for environmental engineers to help address this challenge are highlighted in Box 2-3.

BOX 2-3. EXAMPLE AREAS IN WHICH ENVIRONMENTAL ENGINEERS CAN ADVANCE EFFORTS TO ADAPT TO CLIMATE CHANGE

Environmental engineers, working with civil engineers and experts in climate science and data, can play a number of roles in adapting to the expected impacts of climate change:

Building Disaster Resilience
- Develop a national wildfire smoke forecast system.
- Analyze changing coastal and inland flood risks under climate change and land-use change, including risks to priority infrastructure.

Adapting Urban and Coastal Infrastructure
- Analyze the benefits and costs of gray versus green infrastructure, including pollution control and ecosystem services.
- Identify cost-effective adaptation strategies for water and wastewater infrastructure at risk from sea-level rise.

Ecosystems
- Develop a better understanding of ecosystem services in mitigating the impact of climate change.
- Develop and evaluate approaches to reduce pollutant loading to ecosystems.
- Develop strategies to reduce and mitigate impacts of environmental degradation, deforestation, and ecosystem loss.

Agriculture
- Analyze large-scale costs and benefits of major changes to agriculture, including location and dietary changes.

Health
- Develop sensors capable of rapid pathogen detection in humans, animals, and the environment.
- Use green infrastructure, vegetation, and other methods to reduce urban heat island effects while improving water quality in vulnerable communities.
- Participate in formulation and implementation of innovative strategies to reduce the risk of transmission of vectorborne, zoonotic, foodborne, and waterborne diseases.

GRAND CHALLENGE 3:

Design a Future Without Pollution or Waste

In nature, waste is a resource. One organism's waste is repurposed to sustain another. Since the Industrial Revolution, human society has adopted a more linear model. Resources and energy are used to manufacture products, which are then used and ultimately discarded as waste when those products are no longer wanted (Figure 3-1). This linear model of "take-make-dispose" has been successful in providing affordable products to billions of people and advancing their standard of living. However, this production model generates over a billion tons of discarded products and by-products globally each year (see Box 3-1), and uses large amounts of energy and resources that are never recaptured. An analysis of five high-income countries found that one-half to three-quarters of annual resource inputs are returned to the environment as waste within a year.[159] Despite improved efficiency in the use of resources, the overall production of waste in many countries, including the United States, continues to increase.[160]

The "take-make-dispose" model introduces large amounts of pollutants into the water, soil, and air. Throughout much of the 20th century, large-scale chemical production combined with inappropriate chemical handling and waste disposal created a daunting array of legacy hazardous waste sites globally.[165] Technologies to characterize these sites and contain and remove hazardous contaminants have advanced significantly over the past three decades, and there have been many successes.[166] However, there remain at least 126,000 hazardous waste sites with residual contamination in the United States alone, about 12,000 of which are considered unlikely to be remediated to the point of unrestricted use with current technology. Some of these sites will require monitoring, treatment, and oversight in perpetuity.[167] Meanwhile, new concerns associated with legacy contaminants continue to be discovered (Box 3-2).

NATURAL RESOURCE EXTRACTION

FIGURE 3-1. The linear model of resource extraction, manufacturing, consumption, and disposal ("take-make-dispose") dominates global economies. This model produces ever-increasing amounts of garbage while wasting resources and generating excess pollution.

44 | ENVIRONMENTAL ENGINEERING IN THE 21ST CENTURY: ADDRESSING GRAND CHALLENGES

BOX 3-1. RETHINKING CONSUMPTION AND WASTE

Energy, water, and food resources are routinely wasted along supply chains. For example, food waste includes harvest spillage or damage, losses during processing, and produce thrown away because of blemishes or spoilage (see Challenge 1). For computers and other electronic devices, waste is generated from mining raw materials and from manufacturing processes. Once in use, many end products do not take long to become waste themselves. The plastic sandwich bag that is manufactured from petroleum or the foil wrapper derived from refined bauxite ore are often used for a matter of hours before being discarded. The service life of electronic products continues to shrink due to technical advancement, style preferences, or planned obsolescence.

Globally, about 80 percent of consumer goods, excluding packaging, are disposed after a single use with no plan or ability to be reused, recycled, or biodegraded.[161] Municipal solid waste generated per year is expected to double by 2025[162] and triple by 2100.[163] This upward trajectory is occurring despite increases in recycling and reuse in the developed world primarily because the increasing size of the middle class, which accounts for the bulk of consumer goods spending (see figure below). In 2015, there were more than 3 billion people in the middle class worldwide, and by 2030, the middle class is anticipated to expand by another 2 billion people.[164] Much of this growth is in the developing world, where modern environmentally sound methods to manage waste are less common. Encouraging less consumption, developing product designs and manufacturing that minimizes waste, and increasing recycling and reuse globally is a major opportunity and responsibility of the environmental engineer that will preserve resources for future generations and reduce waste and pollution.

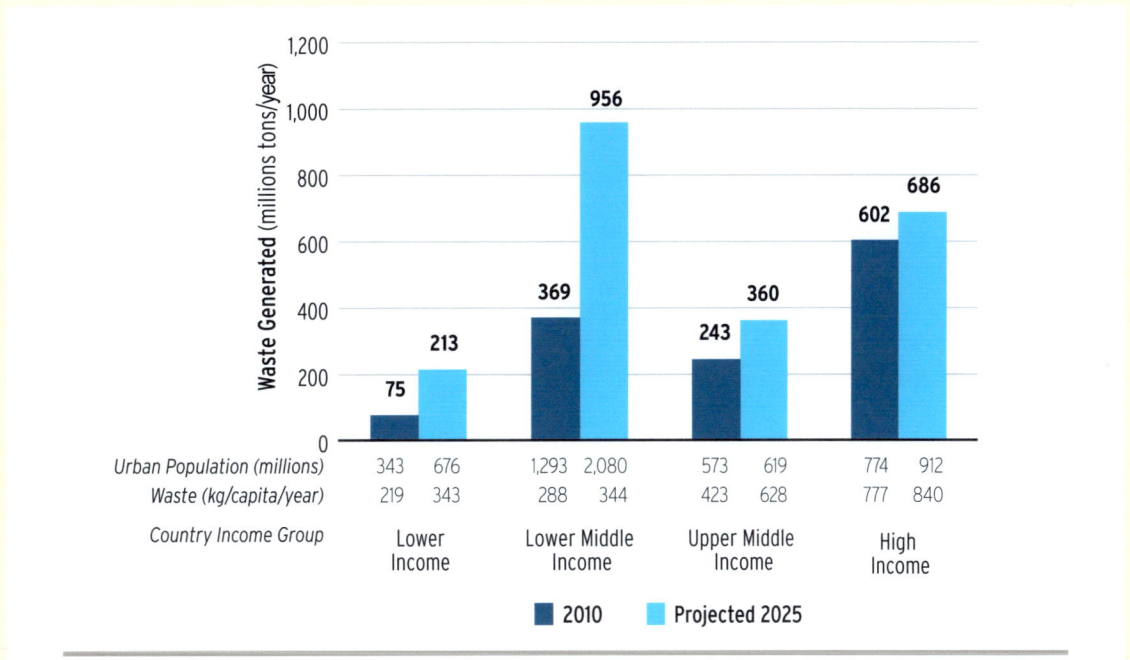

Current and projected urban waste generation by income level and year.

PRODUCTION & CONSUMPTION

WASTE & POLLUTION

Design a Future Without Pollution or Waste | 45

BOX 3-2. EMERGING CHALLENGES WITH LEGACY CONTAMINATION

New concerns associated with legacy contaminants continue to be discovered. For example, per- and polyfluoroalkyl substances (PFAS), which include over 3,000 compounds, have been produced worldwide since the 1940s for use as water-resistant coatings in manufacturing and in fire-fighting foams commonly used at military and civilian airports.[183] Over the past decade, these chemicals, sometimes called "forever chemicals" because they do not biodegrade, have been increasingly detected in surface water and groundwater, sometimes at levels exceeding the U.S. Environmental Protection Agency's (EPA's) lifetime health advisory level (70 ng/L, established based on exposure to two PFAS compounds).[184] Based on EPA sampling of public water supplies in the United States, up to 15 million people live in areas where their drinking water exceeds the EPA health advisory level.[185] However, in mid-2018, the Agency for Toxic Substances and Disease Registry stated in a draft toxicology risk assessment that the EPA level may be 7 to 10 times too high for two common PFAS compounds to protect against health risks.[186] Continued research is needed to determine the scope of the problem, assess the risks posed by the many different chemicals, and develop water treatment options where appropriate to inform policy decisions for use and management of these compounds.

Rapidly developing countries are also facing escalating environmental crises as a consequence of major economic growth without regard for socioenvironmental costs. A key example is China, where industrial development combined with insufficient environmental protection over the past three decades has resulted in widespread soil contamination. China's first national soil survey results are alarming: nearly 20 percent of agricultural lands are classified as polluted.[187] The pollution stems from atmospheric deposition of heavy metals and direct irrigation using industrial wastewater,[188] and human exposure is evidenced by heavy metal contamination in China's rice crop.[189] The scale of environmental cleanup needed to address this problem is similarly alarming, with cost estimates of China's current land remediation plan as much as $69 billion by 2020.[190]

Environmental engineers can help address legacy contamination problems using sustainable remediation approaches. These include stakeholder engagement and life-cycle analysis to identify the best long-term solutions that are socially acceptable and economically viable while minimizing negative side effects of cleanup activities, such as air pollution and ecosystem degradation.[191]

Over the past few decades, the amount of pollution produced by some industries and activities has dropped precipitously thanks to research and technology advances and effective policy interventions (see Challenge 5). For example, regulations on heavy-duty diesel fuel emissions, the development of ultra-low-sulfur diesel fuel, and new emission control technologies have helped reduce particulate matter and nitrogen oxide emissions by more than 90 percent in diesel truck and bus engines put into use since 2010 in the United States.[168] Nevertheless, large quantities of untreated sewage, industrial by-products, and vehicle emissions continue to find their way into the water, soil, and air.[169] Human activities are causing nitrogen and phosphorus to accumulate in bodies of water[170] and greenhouse gases to accumulate in the atmosphere (see Challenge 2).[171] Toxic chemicals have been detected in people and wildlife in every corner of the globe, from the Arctic wilderness to remote tropical islands.[172]

Because of improvements in living conditions, including water treatment, sanitation, and health care, the 20th century saw a doubling of life spans globally,[173] but pollution continues to have profound effects on human health. Pollution contributes

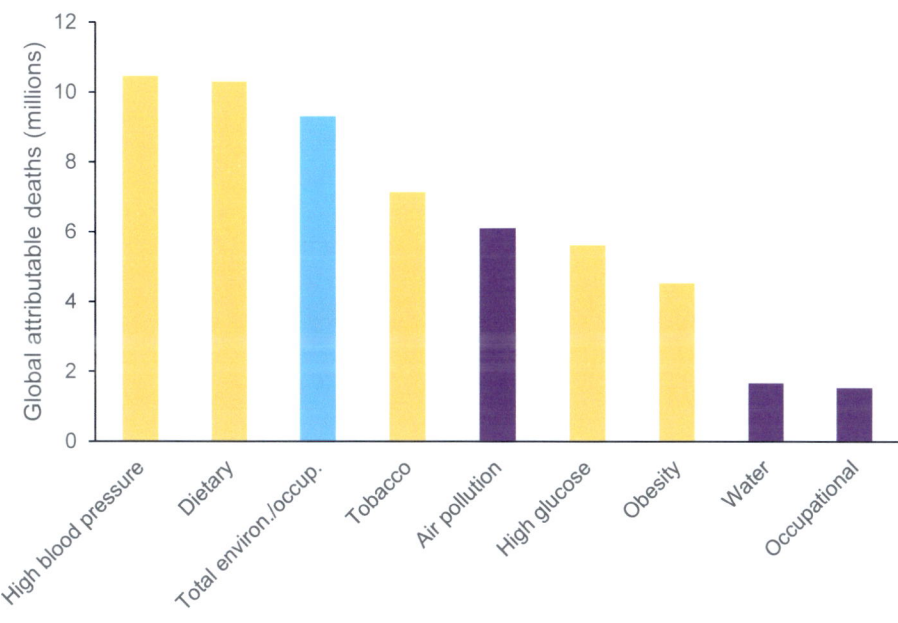

FIGURE 3-2. Global estimated deaths by risk factor and by total environmental and occupational causes (blue), which are disaggregated and shown individually in purple. Air pollution–attributable deaths are primarily linked to particulate matter pollution and indoor burning of solid fuels. Water-related risks are associated with diarrheal disease from unsafe water and poor sanitation. The estimated occupational deaths include 0.33 million from injury, but the remainder are from pollution-related causes, such as asbestos, carcinogens, and airborne particulate matter. The risk factors are not exclusive of one another.

to the leading causes of death worldwide including heart disease, stroke, and chronic lung disease. One of every six deaths in 2015—about 9 million deaths worldwide—can be attributed to disease from exposure to pollution (Figure 3-2).[174] Air pollution causes two-thirds of the premature pollution-related deaths, while unsafe drinking water and sanitation account for nearly 20 percent.[175] More than 90 percent of the world's population lives in areas where air quality does not meet health standards.[176] Although the problems are worse in low- and middle-income countries where the sources of air pollution are minimally controlled, air pollution is estimated to cause nearly 400,000 premature deaths annually in high-income countries.[177] Because these estimates do not account for compounds whose effects are not well characterized, for example, chemicals thought to cause endocrine disruption, the true toll of the health effects of chemicals is likely underestimated.

Pollution also harms natural ecosystems. Metals leaching into streams from abandoned mines have been linked with reduced biodiversity, and trace organic chemicals, such as pharmaceuticals, have been associated with reproductive anomalies including the feminization of male fish.[178] Millions of tons of plastic end up in the oceans every year,[179] creating large floating islands of garbage, and small plastic particles ("microplastics") are accumulating in the food chain with a largely unknown effect.[180] Wastewater discharges, urban and agricultural runoff, and fossil fuel combustion sources have overloaded lakes, estuaries, and rivers with nutrients, fostering algal blooms that can deplete oxygen and produce toxins.[181] All of these ecological problems ultimately harm human health and disrupt industries such as fisheries and agriculture. In 2014, for example, about 500,000 residents of Toledo, Ohio, were ordered not to use their tap water for days due to toxins produced by an algal bloom in Lake Erie.[182]

Challenges posed by pollution and waste will intensify as the world's population grows, people live in ever higher densities, standards of living increase, and industrial production expands to meet increasing demands. Two new approaches will be required to achieve economic progress while minimizing negative health and environmental impacts and sustainably managing Earth's resources. First, a new paradigm of waste management and pollution prevention is needed—one that shifts from a linear model of resource extraction, production, use, disposal, and cleanup toward one designed to prevent waste and pollution from the outset. Second, innovative approaches are needed to recover valuable resources from the waste we do produce. Ideally the two approaches are closely coupled. These new approaches will require life-cycle and systems thinking to identify sustainable solutions that minimize the amount of energy and resources consumed and the amount of waste and pollution generated through all components of production and use.

Preventing Pollution and Waste Through Improved Design

Every day, new chemicals and materials are manufactured, elements are mined from the earth, fuels are burned, and fertilizers and pesticides are made and used. These activities are undertaken to support functions and provide services—such as the production of food, medicines, clothing, building materials, and electronics—that are vital to our society and economy. The question is now how to provide these functions and services without generating the types and scale of pollution and waste that have harmed human health and ecosystems in the past.

The solution requires working toward a circular economy designed to prevent harmful waste and pollution from the outset. Within a circular economy, processes are designed to minimize waste, products and waste materials are reused if possible, and materials that cannot be reused are remanufactured or recycled (Figure 3-3). Organic wastes that cannot be reused are converted to other useful products such as chemicals, materials, or fuels. Pollution prevention is also considered at every design stage to minimize negative impacts. Using materials

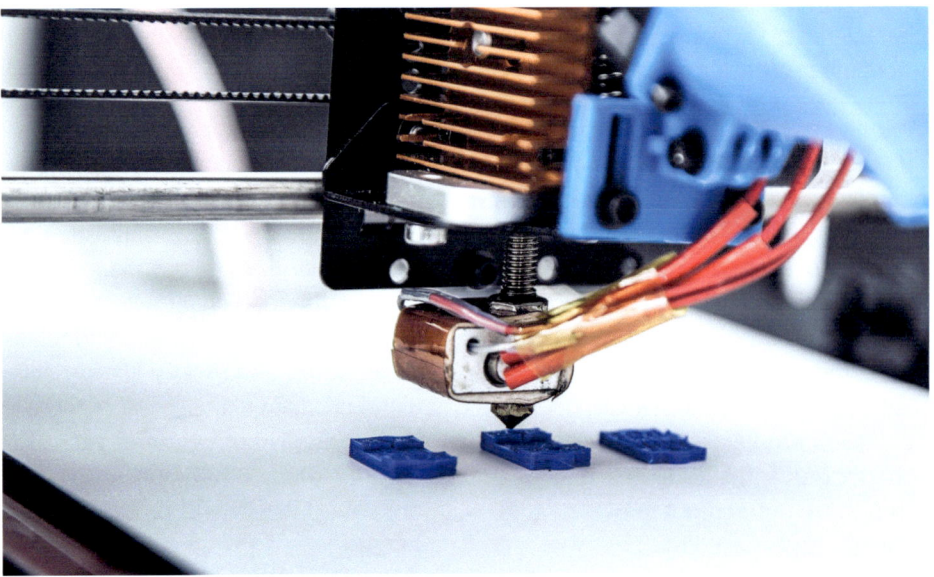

and chemicals that are relatively benign in the environment reduces risks to human and ecosystem health as they are cycled through the economy and society. When considering the entire life cycle, designs that reduce energy use and promote efficiency are emphasized. By thinking beyond incremental improvements (such as treating effluents on site) and working to develop innovative new approaches that eliminate waste and pollution, environmental engineers can help achieve a sustainable future.

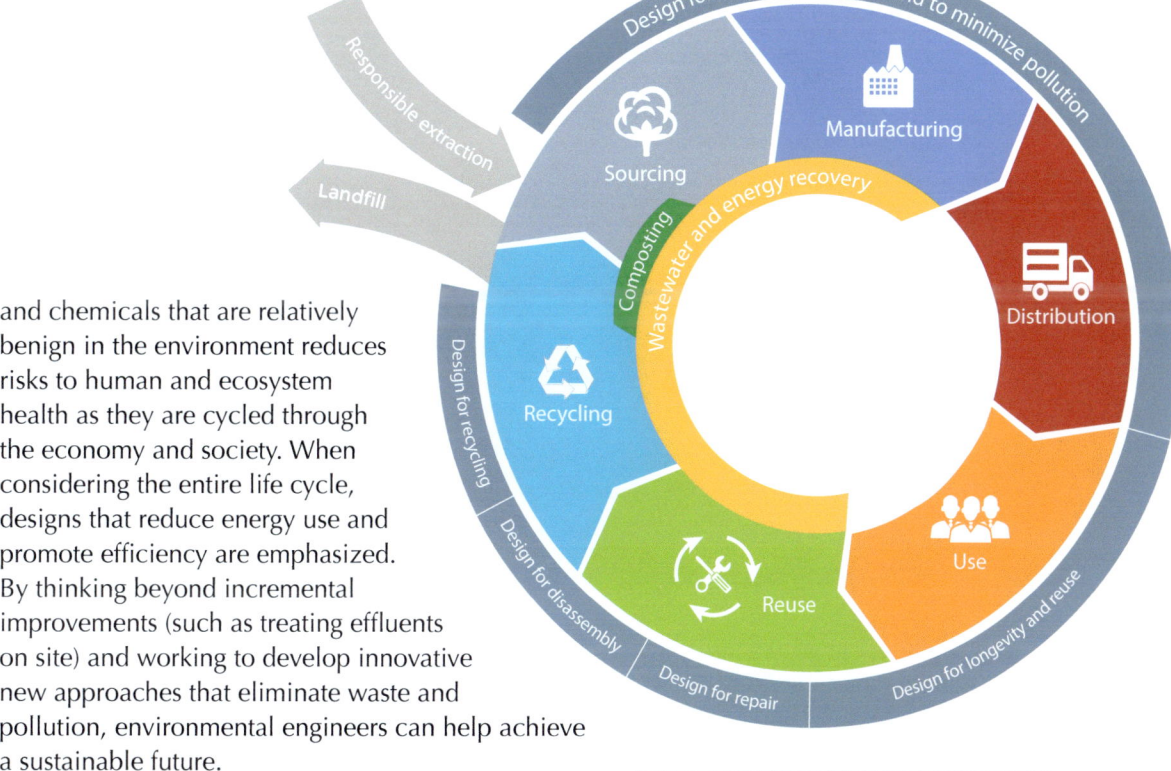

FIGURE 3-3. Sharply reducing waste and pollution requires new approaches to design based on life-cycle thinking.

Design is the stage that most influences the types and amounts of waste or pollution that will be generated. At the design stage, engineers are able to help select and evaluate the characteristics of the final outcomes, considering material, chemical, and energy inputs; effectiveness and efficiency; aesthetics and form; and specifications such as quality, safety, and performance. In the development of new systems, this stage is ideal for innovation and creativity and represents a key opportunity to integrate environmental goals into the specifications of the products or processes. Through life-cycle and systems thinking—as well as green chemistry and green engineering, which emphasize designs that ensure that inputs, outputs, and processes are as inherently nonhazardous as possible—new designs can be implemented that rely on more benign materials and less energy, that do not generate much waste, and that do not shift environmental burdens from one place to another. Benefits of such an integrated approach include wise use of resources, improved human health, and enhanced protection of natural systems. Advances needed to support a circular economy include efficient and effective separation and recycling technologies and market forces or government incentives that recognize the broader impacts of pollution and waste (see Challenge 5).

Many of the most successful interventions focus on preventing the production or release of pollution or waste. This strategy is generally easier and less expensive than remediating contamination sites after toxins are dispersed in the environment. For example, perchloroethylene, a widely used solvent for dry cleaning fabrics and metal degreasing operations and a likely carcinogen,[192] has been replaced in these applications with supercritical carbon dioxide, which has low toxicity and is chemically stable, readily available, and easily recyclable. Another example is the recent movement away from subtractive manufacturing, a process by which three-

dimensional objects are constructed by successively cutting material away from a solid block of material. Instead, additive manufacturing, for example, 3-D printing constructs objects by successively depositing material in layers without the need to generate waste by cutting material away. A growing number of zero-waste businesses and communities aim to reuse, recycle, or recover at least 90 percent of discarded material while also aiming to produce no pollutants to air, water, or land.[193]

Eliminating the use of the most toxic chemicals is an important part of green design. To develop nonpolluting components and processes and prevent future contamination, it will be important to fill knowledge gaps about the full environmental risks of new and existing contaminants. For example, methyl-*tert*-butyl ether (MTBE) was added to gasoline to help reduce emissions in vehicle exhaust. However, MTBE became a groundwater quality problem once gasoline leaked from underground storage tanks because MTBE was able to migrate farther and was more resistant to biodegradation than other compounds in gasoline.[194] Of the more than 140,000 new chemicals that have been introduced since 1950, fewer than half have been subject to human safety or toxicity testing.[195] EPA's Pollution Prevention Framework can be used to estimate physical properties, which are then used to predict environmental concerns such as toxicity, mobility, persistence, and bioaccumulation, but more development and validation is needed. In addition, there are significant needs related to risk communication to help the public and decision makers understand the true costs of pollution.

Capturing the Value of Waste

Under a linear production model, resources are used inefficiently and can become depleted as landfills expand. Recovering resources from waste recaptures the value of those materials and minimizes environmental impacts from further resource extraction. Localized or distributed recovery and reuse also reduces the energy requirements and pollution associated with transportation of materials and waste. Resource recovery can also address local resource shortages in economically depressed or geographically isolated communities.

Today's common waste recovery efforts focus on recycling plastic, glass, paper, aluminum, and scrap metals, but much more is possible with advances in engineered environmental processes that allow the extraction of specific components from waste mixtures. Precious and rare-earth metals could be retrieved from electronic waste[196] and potentially even mined from landfills. Carbon capture systems could be used to turn carbon dioxide into forms that are useful for applications ranging from building materials to plastics to greener solvents.[197] Nutrients in wastewater could be captured for use as fertilizers (see Box 3-3).

Many of today's municipal and agricultural waste streams are rich in organic carbon, which could be recovered and channeled toward chemical manufacturing or energy recovery.[198] The amount of energy contained in wastewater is equivalent to several multiples of the amount of energy required to treat it.[199] Energy recovery has been implemented at numerous centralized wastewater treatment plants, including in Oakland, California, and in Strass, Austria, by converting a fraction of the incoming organic carbon to biogas to produce heat and electricity.[200] However, technologies have not yet been developed to cost-effectively capture the full potential of the embedded energy.[201]

Recovery of resources from waste streams has long been practiced, but in a nonsystematic fashion. In Dharavi, India, one of the largest slums in the world, people have built a thriving economy, employing approximately 250,000 people, based on recovering waste generated in Mumbai. "Gobar gas," produced from anaerobic digestion of animal waste, is used for cooking and community-scale lighting in rural and urban communities, particularly in Southeast Asia and sub-Saharan Africa. Fly ash and gypsum by-products of coal combustion have been used in the manufacturing of concrete and wallboard.[207]

> **BOX 3-3. NUTRIENT RECOVERY**
>
> Nutrients present in wastewater can cause problems for the environment and infrastructure, such as algal blooms in lakes and estuaries and buildup of the mineral struvite in the mechanical systems of wastewater treatment plants. Globally, humans release about 30 percent more phosphorus and twice as much nitrogen into the environment, mostly from fertilizers, than aquatic ecosystems can bear without degrading habitats.[202] Reusing nutrients in existing waste streams can help mitigate these challenges while producing valuable services. For example, reuse of municipal wastewater or agricultural runoff for irrigation can reduce fertilizer use.
>
> Innovative approaches to cost-effectively recover and reuse nitrogen and phosphorus from waste streams rather than mining new phosphorus or synthesizing new nitrogen could conserve natural resources, reduce pollution, and save energy. Phosphorus is an increasingly scarce natural resource with limited mineable reserves,[203] but the phosphorus available from human urine and feces could account for 22 percent of the global phosphorus demand.[204] Recovering phosphorus from waste thus helps to preserve phosphorus reserves for the future, but further advances in waste separation are needed to achieve the technical and economic viability for widespread adoption.[205] In addition, producing reactive nitrogen for fertilizer from inert nitrogen gas in the atmosphere requires a considerable amount of energy and creates further imbalance in the global nitrogen cycle.[206] Some wastewater facilities have been successful in extracting phosphorus to create a commercial fertilizer, but in most cases, recovery of phosphorus and nitrogen from wastewater using *current* technologies is not economically viable.
>
>

Resource recovery is rapidly being integrated into existing manufacturing, agricultural, and industrial practices, but much work remains to be done to realize its potential, both in terms of recovery yields and the types of resources that can be cost-effectively recovered. A significant impediment to utilization of waste is that existing traditional waste streams have not been systematically characterized with resource recovery in mind. Local, regional, and global inventories of waste materials are needed to identify opportunities for reuse or inputs to other production schemes. With this information, appropriate technologies for resource recovery can then be developed using physical, chemical, and biological processes that capture the maximum financial, social, and environmental benefits. This information could also lead industries to redesign their resource extraction and manufacturing processes to reduce waste and more efficiently and cost-effectively recover and reuse valuable resources.

Results of public programs to reduce, reuse, and recycle have been mixed. The United States, for instance, recycles or composts 35 percent of its municipal waste and less than 10 percent of its plastics,[208] but higher rates are possible. Six countries recycle or compost more than half of their waste, led by Germany at 65 percent and South Korea at 59 percent.[209] In 2016, nearly 48 million metric tons of electronic waste were produced globally, representing a value of approximately $60 billion in raw materials, and only 20 percent of this waste was recycled.[210] EPA reports that electronic waste accounts for 70 percent of heavy metals in landfills, such as mercury, lead, and cadmium.[211] Waste streams are often heterogeneous, complex mixtures that currently require significant resources and energy to separate. Sorting technology has been developed and commercialized for some wastes, such as separating organic from inorganic wastes. The extent of resource recovery from wastes could be enhanced by improved, cost-effective waste separation techniques.[212]

Effective waste recovery requires attention not only to scientific and engineering capabilities but also to economic and behavioral factors. Considerations of financial viability and feasibility include the cost of the recovery technology, the quality of the recovered product, the market for the product, any adverse environmental impacts, and measures required to manage and prevent them. Governments can also develop incentives to encourage waste recovery that account for broad societal and environmental benefits of these programs (see Challenge 5).

Many of these advances are focused on large urban areas, where the highest volumes of waste are generated. However, there is also substantial potential to harvest the value of waste streams that are smaller or more intermittent to benefit rural communities. For example, decentralized resource recovery systems could be developed, particularly for sewage, food, animal, and agricultural waste.

What Environmental Engineers Can Do

With training in environmental chemistry, microbiology, hydrology, transport processes, solid waste management, water and wastewater treatment, and air pollution—as well as skills in life-cycle and systems thinking—environmental engineers bring important capabilities toward designing a future without pollution and waste (see Box 3-4). Technological advances combined with innovative new materials and designs can be used to conserve natural resources and minimize adverse effects on human health and the environment. These complex challenges demand solutions that consider broad costs and benefits throughout the life cycle, including human health risks, environmental impacts to water, soil, and air, as well as social and financial impacts (see Challenge 5). Environmental engineers can help analyze the impacts of innovative manufacturing and resource recovery approaches compared to the life-cycle impacts of traditional processes to identify the most promising solutions.

For many pollutants, although the knowledge and technology exist to reduce exposure, the greater challenges are economic, political, and social. For example, billions of people worldwide use solid fuel–burning cookstoves for daily meal preparation, creating large amounts of particulate matter pollution. It is possible to design cookstoves that are much cleaner burning to benefit health, local environmental quality, and climate, but there are cultural, economic, and logistical hurdles to their adoption.[213] Improving resource recovery in developed countries may require people to change their behaviors and accept new approaches to waste separation. An interdisciplinary approach applying social and cultural knowledge is crucial to overcoming such hurdles to guide the development and adoption of sustainable solutions.

BOX 3-4. EXAMPLE ROLES FOR ENVIRONMENTAL ENGINEERS TO DESIGN A FUTURE WITHOUT POLLUTION OR WASTE

Environmental engineers have essential skills needed to move toward a future without pollution or waste. Examples of ways environmental engineers can contribute include

Preventing Pollution and Waste
- Redesign products and their production processes to promote resource efficiency, longevity, reuse, repair, and recycling while minimizing pollution.
- Develop and use tools to better predict the risks of new and existing chemicals in the environment, including toxicity, fate, and transport.
- Quantify and document the life-cycle consequences associated with producing commonly used resources and products and the broad costs and benefits of alternative approaches designed to reduce pollution and waste. Work with social and behavioral scientists to communicate this information to inform the decisions of consumers, manufacturers, and governments that could incentivize these efforts.
- Manage or remediate existing legacy hazardous waste and contaminated sites to eliminate harmful exposures and return sites to productive use.

Capturing the Value of Waste
- Quantify waste-stream characteristics and identify opportunities to reuse or recover materials traditionally considered as waste.
- Identify products that could be manufactured with recycled and reused materials that would have lower cost, lower greenhouse gas emissions, and require less energy to produce.
- Develop new resource-recovery technologies and processes for cost-effective recovery of materials and energy from the waste stream.
- Work with other sectors including public health, architecture, and urban planning to integrate engineering designs, processes, and technologies to develop effective approaches to resource recovery with broad societal benefits.

GRAND CHALLENGE 4:

Create Efficient, Healthy, Resilient Cities

The future is increasingly urban. Cities will absorb almost all of the world's projected population growth in the next three decades. By 2050 cities will be home to over 2 billion *more* people than today. The proportion of the world's population that lives in urban areas will grow from 55 percent in 2017 to 66 percent in 2050.[214] By 2030, 10 more cities are expected to cross the 10-million-inhabitant threshold for the first time, increasing the number of "megacities" from 31 in 2016 to 41 in 2030. The majority of these will be in lower-income countries and contain large slums—dense informal developments without government services.[215]

While this massive concentrated population growth is likely to further compound many of the current problems that cities face, the urbanization of the human population is happening at least in part because of the inherent attractiveness of cities. They offer significant educational, economic, and cultural opportunities as well as better access to communication and health care services. These opportunities draw migrants from the rural countryside where such opportunities are sparser. As noted in a 2016 United Nations report on urbanization, cities are seen as economic hubs and drivers of innovation and competition, propelling a steady flow of people from rural to urban areas, particularly in Asia.[216]

Even as this economic attraction accelerates urbanization, today's cities face persistent problems associated with air and water pollution, energy distribution, water supply, waste disposal, and waste generation. Although cities only occupy 3 percent of Earth's ice-free landmass, they produce 50 percent of global waste and 60 to 80 percent of global greenhouse gas emissions. Cities account for 60 to 80 percent of the world's energy use and 75 percent of all natural resource use.[217]

Cities have stark inequities in the distribution of incomes, public services, access to open space, and quality of life. In middle- to high-income countries, urban sprawl and car-centric and inefficient transit systems create traffic congestion, pollution, and safety hazards, degrading quality of life. Lack of green space and abandoned properties contribute to social and environmental stress, especially in poor urban neighborhoods. Urban communities are fractured by poverty and unequal access to community services, even as accelerating gentrification exacerbates those inequities.

In low- and middle-income countries, large populations live in dense informal settlements that are expanding rapidly; about 880 million people live in slums today and that number is projected to more than double by 2050.[218] With many cities unable to provide adequate sanitation or food and water security for these slums, their residents face a high risk of malnutrition and disease.[219] Increased human contact with domestic animals and wildlife in these settings heightens the risk of diseases with pandemic potential that emerge from animals and subsequently spread from person to person, as occurred with the SARS epidemic. SARS spread rapidly to more than 30 counties before being contained.[220]

The functioning and stability of many of the world's major cities are made all the more precarious by threats from extreme events such as floods, heat waves, and droughts, which are expected to hit cities harder and more frequently in the coming decades, putting more lives and infrastructure at risk.[221]

These challenges, however, are not insurmountable. The scale and structure of cities offer unique opportunities to improve quality of life and equitably address many of the grand challenges such as climate change adaptation, pollution, waste, and sustainable food, water, and energy supplies. Aging physical infrastructure represents both a major challenge and a key opportunity to reshape tomorrow's world. The American Society of Civil Engineers has estimated that $4.6 trillion in U.S. infrastructure investment will be needed by 2025,[222] and the Organisation for Economic Co-operation and Development estimates worldwide infrastructure needs at $70 trillion by 2030.[223] If this infrastructure were refashioned to support multiple city functions and the lives of residents in an integrated way, it is possible to create cities that are more equitable, efficient, healthy, and resilient. Environmental engineers can bring unique training and analytical skills to build partnerships with the other professions—in planning, energy, and transportation, among others—who together can creatively overcome these challenges and take advantage of the significant opportunities that cities present.

What Does an Efficient City Look Like?

Cities can be viewed as urban ecosystems, composed of systems of infrastructure networks (such as water, energy, transportation, waste, and public spaces), the people who use and operate the infrastructure, and the multiple interactions that occur between them. Accordingly, urban infrastructure is a system of systems through which energy, money, information, and materials flow. Significant inequity in the distribution of resources and political power within cities can result in infrastructure systems that serve different communities to different degrees.

There are multiple ways to make cities more efficient, both by increasing the efficiency of their individual parts and by making various systems function more in concert with each other. For example, waste from one system can be used in another system (waste to market or waste to energy), thereby minimizing inputs and reducing net waste (see also Challenge 3). Documenting inequities in the distribution of infrastructure services can help urban planners and engineers work to address those issues. Two approaches to improve a city's efficiency involve reenvisioning urban infrastructure and incorporating smart systems.

Reenvisioning Urban Infrastructure

Cities cannot achieve these desired efficiencies by simply monitoring and improving the operations of older infrastructure. In the past, infrastructure systems were designed to optimize water delivery, energy provision, transit, and land use in a siloed fashion that led to suboptimal solutions. Going forward, sustainable urban infrastructure development needs to look beyond the local scale and consider transboundary infrastructures across regional, national, and global scales.[224] For example, developing reliable, nutritious, and sustainable food supplies in densely populated cities requires looking beyond a city's boundaries to the full range of producers, suppliers, and transporters and the implications to energy and water use and greenhouse gas emissions.

Cities can be more efficient by considering the urban infrastructure as a system of systems at many scales rather than individual disconnected entities (energy, water, sanitation, and traffic). The design of buildings and communities affects how much energy and water are used and how much waste is produced (Figure 4-1). Low-impact development that mimics natural processes, for example rain gardens and porous pavement, reduces uncontrolled stormwater runoff and its associated water pollution and erosion.[225] It also provides additional benefits such as added urban green space, reduced urban heat island effects, and recreation opportunities. Improved management of urban stormwater runoff increases nutrient and organic matter concentrations in wastewater, making it easier to recover valuable resources, such as energy and nutrients. Urban aquaponic systems, in which fish and plants are grown together, can recycle wastes and nutrients while providing food security and eliminating food deserts.

Integrated urban solutions that address multiple needs or challenges can also help save money. For example, in lieu of filtration to maintain water quality control

FIGURE 4-1. The Bullitt Center in Seattle is an example of a building designed to minimize its environmental impacts. Constructed of local materials selected for their low health and environmental impacts, the building has solar panels that generate as much energy as the building uses, employs geothermal heating and cooling, actively controls windows and shades to optimize natural lighting and circulation of fresh air, stores rainwater for nonpotable use, and has its own wetland to filter graywater.

of pathogens, for about 90 percent of its supply New York City uses watershed protection strategies combined with chlorination and ultraviolet disinfection. This approach up to the present time has allowed the city to save $8 billion to $10 billion in capital expenses and approximately $1 million per day in operational costs as compared to an engineered filtration-based approach for the entire supply.[226]

There are significant opportunities to transform urban infrastructure, but also large challenges: Most of the residential and commercial buildings and other infrastructure in today's cities are old and inefficient, needing significant investment to be maintained let alone enhanced. Older infrastructure is especially prevalent in the poorest urban communities, further exacerbating inequities. This transformation to efficient, sustainable urban infrastructure—and the contributions of environmental engineers to that transformation—will need to address head-on how to apply those changes not just to new buildings and infrastructure, but to adaptive reuse and revitalization in all city neighborhoods.

Advancing Smart Cities

Improvements in efficiency can also be gained through "smart" technologies that capitalize on advances in sensing technology, data, connectivity, artificial intelligence, and participatory governance to optimize operations and resource management.[227] A smart system can be not only reactive but proactive, using inputs, information processing, intelligence, and actuation to anticipate and prevent

> **BOX 4-1. DESIGNING SMART SYSTEMS**
>
> Smart systems are being developed to improve city functions, as demonstrated by these examples. Such systems can become more predictive and require less human involvement as technology improves and smart systems are more effectively integrated into city operations.
>
> - A start-up company is applying artificial intelligence to help cities efficiently respond to earthquakes by predicting, in real time, which areas are likely to have suffered the most damage and where injured people are likely to be concentrated.[231]
> - In Barcelona, sensors provide site-specific weather information that is used to calibrate the precise amount of water needed to irrigate parks.[232]
> - In Amsterdam, a mobile app allows cyclists to turn up the intensity of outdoor lighting while they ride along a bike path and then let the lights dim after they pass through, allowing residents to play an active role in helping the city operate efficiently.[233]
> - Smart grid technology in being implemented to improve efficiency and avoid cascading failure, as happened in the 1996 western United States blackout.[234]
> - Smart waste management systems monitor how full bins are and use solar power to compress waste before pick-up, helping managers plan waste collection routes for greater efficiency.[235]
>
> Smart systems are not limited to cities in higher-income countries; low- and middle-income countries have begun to "leapfrog" over older technologies to take advantage of newer ones:
>
> - A collaboration between the World Bank, the ride-hailing platform Grab, and the government of Cebu City, the Philippines, allows the city government to use GPS data from taxi drivers' smartphones to track traffic patterns and incidents and inform emergency response and transportation planning.[236]
> - An app developed for residents in Rio de Janeiro, Brazil, uses crime data and machine learning to predict where and when crimes are most likely to occur, allowing users to make informed decisions and reduce their risk as they move throughout the city.[237]

problems or inefficiencies.[228] Although there are many ways to define a "smart city," the basic idea is that cities can improve outcomes, such as efficiency or quality of life, by incorporating smart interconnected systems into municipal functions.[229]

Technological advancements are increasing opportunities to develop smarter cities. Improvements in sensing technology have made it feasible to collect detailed geospatial and other types of data on the systems that keep cities ticking, such as transportation patterns and water and energy use. When appropriately analyzed and connected to decision making or operational controls, these data can be a powerful asset to improve city functions and planning. Developments in data science and machine learning are advancing these capabilities; a 2018 report by the World Economic Forum[230] identified artificial intelligence as a key technology for efforts to transform traditional sectors and systems to address climate change, deliver food and water, protect biodiversity, and bolster human well-being.

Smart systems are being tested in cities around the world. To date, most of these tests focus on isolated sectors, such as transportation, emergency response, or electricity distribution (see Box 4-1), although some projects are experimenting with combining multiple smart systems across a community (see Sidebar).

Despite these encouraging developments, adaptive, full-scale predictive smart cities are still a long way off. Questions around performance, control, security, economics, equity, and ethics in smart cities must be addressed in order to fully realize the complete suite of societal benefits.[241] In addition, while sensors and

DEVELOPING SMART COMMUNITIES

Smart communities that combine multiple smart systems are being planned around the world. One such planning effort is under way in Toronto in a formerly industrialized area along the waterfront, called Quayside. Sidewalk Labs, a subsidiary of Alphabet, is planning to take a 12-acre plot and create the "world's first neighborhood built from the Internet up,"[238] with flexible mixed-use space and housing for about 5,000 people (see figure below).[239]

Some of the smart features envisioned at Quayside include a carbon-neutral thermal grid that would use geothermal energy, waste heat, and energy generated by anaerobic digestion of organic waste to heat and cool buildings, combined with rigorous building construction standards aimed to reduce energy demands. Autonomous shuttles, cycling, and walking would be the primary means of transit. Sidewalk snow melters and automated awnings would keep bike-share stations, transit stops, and cycling and walking paths useable through the winter.

Sensor deployment and data acquisition represent the backbone of the Quayside project vision. Sensors would measure everything from air pollution and noise to sewage flow rates to how often a public waste bin is used.[240] The design process was launched in 2017 and Sidewalk Toronto is working with experts and stakeholders to co-create the final neighborhood design plans.

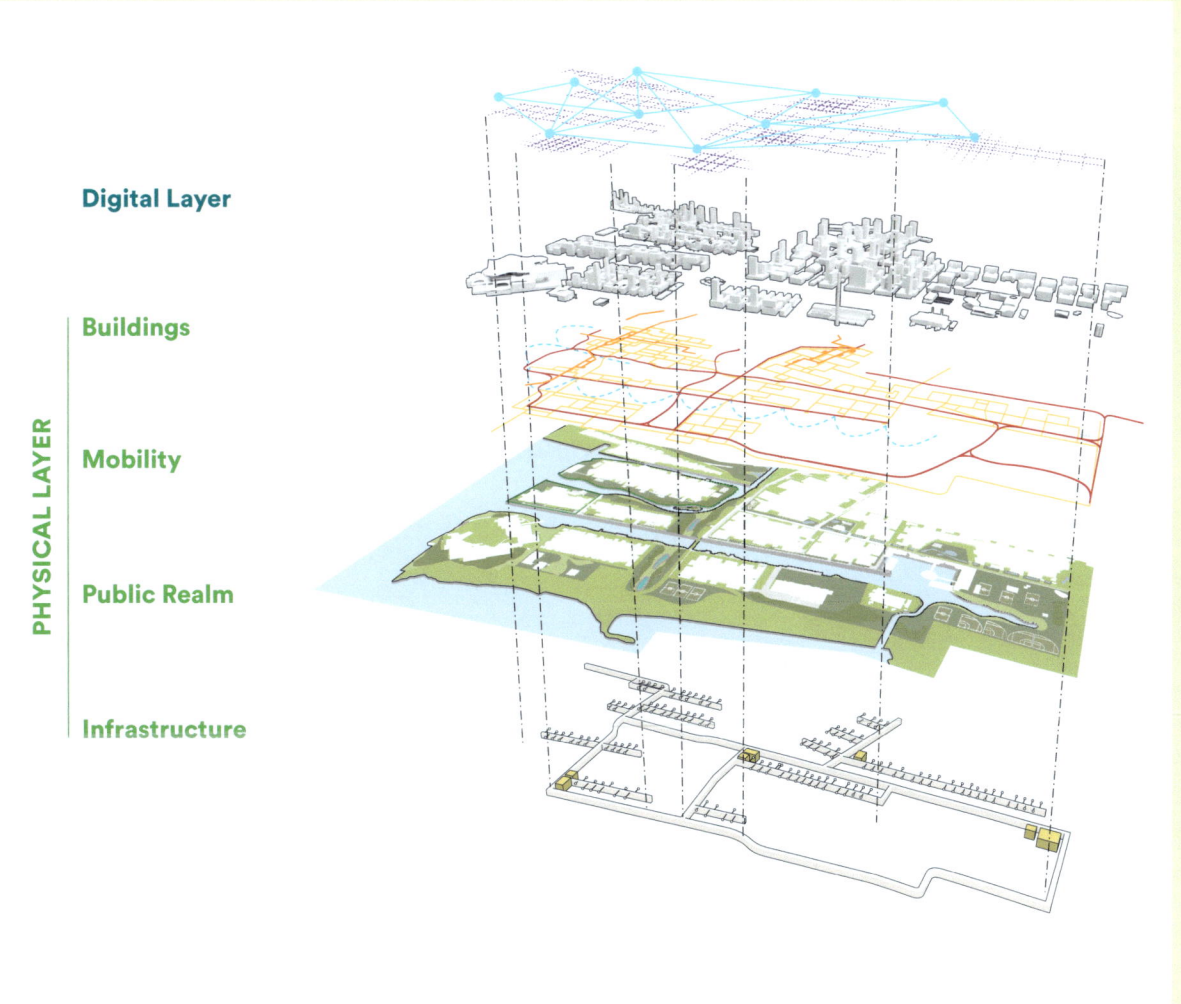

A vision for Quayside, a mixed-use urban development in Toronto. The design process was launched in 2017.

tools for citizen participation are being actively developed, implementation—effectively integrating real-time information and feedback into actual city operations—remains a challenge.[242] These challenges are not mere technological hurdles; overcoming them will require deep understanding of the physical and social systems that are integral to city functions, as well as a deeper understanding, on the part of engineers and city managers, of the opportunities and limitations of the technology. Furthermore, there is a need not only to continue to develop and scale tools for collecting data, but also to facilitate the effective and equitable application of such information. This will require interdisciplinary efforts to manage and interpret data, a willingness and capacity to adapt city operations to changing circumstances, and adequate protections for privacy and security.

What Makes a City Healthy?

Healthy cities facilitate good health and promote a high quality of life for all their residents. Healthy cities support mental and physical health, providing residents sufficient and equitable access to community services, education, housing, art, clean rivers, recreation, and green space, as well as protection from crime, violence, and hazardous environments. Clean air, safe drinking water and sanitation, effective and affordable transportation, reliable access to power, ample opportunities for employment, and access to nutritious food and health services are important facets of a healthy city.

Healthy buildings are a critical component of healthy cities because people spend over 90 percent of their time indoors.[243] Healthy buildings are constructed of materials that do not off-gas toxic compounds into the air. They feature ventilation and lighting designed to optimize productivity and well-being, while also conserving energy. Designing buildings for health, well-being, and water and energy conservation can sometimes involve trade-offs to optimize for competing needs. For example, a tightly sealed building is more energy efficient with respect to temperature control, but it also allows build-up of contaminants in air. Likewise, technologies and practices designed to save water and reduce energy used for heating water may inadvertently promote the spread of pathogenic microbes.[244]

The capacity to prevent, detect, and mitigate the spread of infectious disease is particularly vital to healthy cities, but it will become more difficult to establish and maintain this capacity as cities and slums grow larger and denser. Many emerging diseases result from transmission of infectious agents from animals to people.[245] These trends underscore the need to take a holistic view of public health that encompasses the health of humans, animals, and the environment, a concept and approach known as One Health.[246] Two important infectious disease challenges are the emergence of diseases with pandemic potential and the emergence of antibiotic-resistant pathogens.[247] Although these challenges are not uniquely urban, many infectious disease problems could be exacerbated in cities and spread through connected suburban communities.

Sophisticated techniques such as culture-independent diagnostics, genomic analysis, and advanced epidemic modeling offer valuable tools to track and contain the spread of pathogens and antibiotic-resistant organisms. Yet, these tools cannot make up for a lack of basic infrastructure to deliver clean air, safe food and water, sanitation services, and reliable electricity to homes and health care facilities. The knowledge and technology to mitigate many of the environmental drivers of infectious disease and other public health threats exist, but there are significant gaps in infrastructure and services, especially in the poorest areas. This disparity points to a need for more efficient, scalable solutions to support public health, including measures to prevent and contain infectious diseases along with improvements to the broader social and physical environments of the world's cities.

Innovative solutions have been proposed to apply technologies and policies to improve public health in low-income settings. In Africa's largest urban slum (Kibera in Nairobi, Kenya), integrated "biocenters" are being used to capture waste and digest it into biogas, which can be used as cooking fuel, thereby helping to manage waste while simultaneously reducing exposures to both outdoor and indoor air pollution from traditional cooking with wood, dung, and charcoal.[248] The Diesel Emissions Reduction Act[249] has provided grant funding and other incentives to support clean diesel projects, helping to replace old diesel school buses in low-income communities in Houston with low-emission models, reducing children's exposure to pollution from diesel exhaust. With new emission standards and advances in technology, the percentage of low-income populations in the United States that live with air quality above the current fine-particulate standards dropped from 57 percent in 2006-2008 to 8 percent in 2014-2016.[250]

What Makes a City Resilient?

Resilient cities have the capacity to endure disasters, whether they are economic, environmental (such as floods, earthquakes, or drought), or the result of terrorism. To be resilient, cities must have the ability to withstand stress and quickly recover or adapt. One way to accommodate stress is to have redundant systems, for example, in utilities such as power or water grids or transportation routes, to support continued operations when the primary system is not functional. Resilience also means being able to mobilize resources quickly in response to a disruptive event and contain the amount of damage caused. Resilience encompasses preparation, response, recovery, and adaptation.

Increasing a community's resilience can involve repurposing existing systems or creating infrastructure that serves multiple purposes. Boston's Muddy River Restoration Project restores riparian habitat to reduce the severity of flooding events.[251] The project discourages development in flood-prone areas, reducing the damage, displacement, and disruption associated with future floods. In the wake of significant flooding, Copenhagen's Østerbro neighborhood is creating a network of green streets and neighborhood park stormwater retention areas that will make the neighborhood more resilient to future storms.[252]

To increase resilience, it is important to systematically assess current vulnerabilities to inform better design. Such assessments can be used to prioritize measures for addressing vulnerabilities through existing and planned systems and infrastructure. Climate science provides one input into such an assessment. For example, planners can use decision tools to examine the range of potential infrastructure impacts associated with future climate scenarios, projecting threats such as sea-level rise, drought, and extreme heat. Planners also need to look at anticipated shifts or stressors that are likely to affect a city's ability to respond to such events. For example, it may be important to assess transportation patterns and factors that may influence the number and use of vehicles. As a city's population rises, dramatic increases in the number of vehicles could overwhelm infrastructure and necessitate the replacement of open lands with parking areas and buildings that would exacerbate flooding and increase heat island effects. On the other hand, a significant increase in the use of shared vehicles, as might occur with autonomous vehicles, could eliminate 90 percent of parking demand,[253] thereby reducing projected flood risk and allowing the repurposing of parking space for the creation of green space.

Many cities are actively pursuing sustainable, multipurpose solutions like those in Copenhagen and Boston, but the scale of these projects is often not aligned

with the full scale of the challenge, and there remains much room for improvement toward understanding risks and building more resilient structures, systems, and communities. In addition to research and technological solutions, becoming resilient requires a cultural shift among decision makers, stakeholders, and citizens. By better assessing, understanding, and communicating risk, cities can garner support for forward-looking resilience goals and the steps needed to achieve them.

What Can Environmental Engineers Do?

In general, efficient, healthy, resilient cities will not be built from scratch. Rather, the challenge is to incorporate new designs and systems into existing cities and their infrastructure. This means actively reengineering existing land-use patterns, built environments, and water, sewer, electricity, and transportation modalities and infrastructure. What's more, cities must undertake these efforts at the same time as they are absorbing massive population growth, that stresses current systems as new ones are established. This will undoubtedly be a complex process, and implementing effective solutions will require research and coordination involving multiple disciplines and sectors. Research is needed to identify and prioritize key vulnerabilities that cities face and effective adaptations that they should undertake. These efforts should include gleaning lessons from cities that have begun such transitions, as well as finding innovative ways to engage both the private sector, which has significant sway over the state of the built environment, and the public sector, which typically leads the way on infrastructure.[254]

Creating efficient, healthy, resilient cities involves many overlapping considerations from the challenges discussed previously in this report. The solutions will require leadership, systems thinking, and innovation from environmental engineers working with the many other professionals—in planning, transportation, energy, and public health, among others—to create and implement successful urban solutions. In particular, the tools of environmental engineering will be invaluable in applying sensors strategically, building distributed systems, and improving the design of cities.

Applying Sensors Strategically

Sensors are key to smart, responsive cities and are particularly valuable for conserving resources and increasing livability and safety. Traffic-monitoring sensors, for example, can be used to change signal patterns to relieve congestion in real time or inform long-term solutions to more systemic traffic issues, thus reducing the amount of energy wasted, pollution generated, and productivity lost in traffic jams. Similarly, sensors that collect data on water or energy use can help individuals minimize their consumption of these resources and inform how utility companies manage and deliver them or respond to disruptions. Systems that monitor chemical or biological contaminants in air, water, food, and human populations can provide early warning of emerging health threats. Sensor technology is developing rapidly and in many cases is now good enough and cheap enough for widespread use;

the question is, how can these technologies be deployed and utilized through applications of artificial intelligence algorithms to enable efficient operations at the scale of a city?

Building Distributed Systems

Although many of today's cities are built with centralized systems for water, energy, and waste, distributed systems could make cities both more efficient and more resilient. For example, buildings or city blocks can generate their own electricity by incorporating renewable sources such as solar, wind, biomass, or wastewater. Or, they can reduce their reliance on centralized water supplies by collecting graywater, rainwater, or cooling water and using it for nonpotable purposes.[255] Multimodal systems, such as combined cooling, heating, and power systems, use the waste heat from electricity generation to heat or cool buildings; these systems can be twice as efficient as separate systems[256] and also reduce greenhouse gas emissions, air pollutants, and water consumption.[257] A city with a combination of centralized and distributed systems also means that, in a time of disaster, people live closer to the services they need and are less heavily impacted by disruptions that occur elsewhere in the city. These same distributed systems could also be customized for use in rural areas, providing access to services that are costly to deploy in areas with low population density.

Although there are now many emerging technologies and models to support distributed systems, environmental engineering expertise is needed to determine which solutions are most practical, resource efficient, and appropriate for different circumstances and to optimally integrate these solutions into existing city infrastructure. At the same time, it is important to continue to develop, optimize, and apply distributed solutions to address the anticipated demands and needs of future cities. To ensure that these solutions are practical and palatable for communities, environmental engineers will also need to be trained to look beyond the technology opportunities and understand perceived and real unintended impacts, such as noise and emissions, which have stymied previous efforts to distribute energy generation in cities.

Improving City Design

Revising the design of cities will be necessary to accommodate more people in a way that improves rather than harms quality of life. Connectivity is one important

element. Connecting people from all economic strata to basic goods and services—from clean water and reliable electricity to groceries and health care to employment—improves equity, health, and resilience. Improving infrastructure for active transportation (walking and cycling) can enhance health and reduce congestion, energy use, and pollution. Optimizing the design of buildings and public spaces—and identifying ways to build those principles into the revitalization of existing buildings and public spaces—is another key goal that can reduce resource consumption and improve environmental quality and quality of life.

Adapting to climate-related changes and designing for resilience will be key to sustaining cities and their populations in the coming decades. Such adaptations often can serve multiple purposes; for example, equitably distributed green space can promote well-being,[258] while also mitigating natural disasters by absorbing floodwaters and recharging aquifers. Stakeholder engagement is paramount to ensure citizen support for new city designs (see Challenge 5). By identifying, prioritizing, and implementing solutions that will reap multiple benefits, environmental engineers can make significant contributions toward building more efficient, healthier, and more resilient cities. Box 4-3 provides specific examples of ways that environmental engineers can work to create efficient, healthy, resilient cities.

BOX 4-3. EXAMPLE ROLES FOR ENVIRONMENTAL ENGINEERS TO CREATE EFFICIENT, HEALTHY, RESILIENT CITIES

The following are examples of ways that environmental engineers, working collaboratively with other disciplines, can engage with the public and private sectors to help build efficient, healthy, resilient cities. In doing so, environmental engineers can ensure as well that solutions like those highlighted below are designed and implemented in ways that are fully cognizant of—and help to address—the significant current inequitable distribution of services in today's cities.

- Design and revitalize infrastructure systems, including water, energy, food, buildings, parks, and transportation systems, to achieve equitable access and optimize among sometimes competing objectives for health, well-being, water and energy conservation, and resilience.
- Evaluate the potential positive and negative consequences from alternative infrastructure designs, including impacts to pollution, energy consumption, and greenhouse gas emissions.
- Address extraordinary infrastructure challenges in low-income country settings by developing and evaluating innovative approaches to address water, sanitation, and health challenges unique to urban and periurban slums.
- Identify opportunities in cities and design systems for capturing and repurposing waste (solid waste, wastewater, and heat) for energy or resource recovery, considering both large, centralized and small, decentralized systems.
- Develop and use sensors to support more efficient city operations, including transportation, water and wastewater, energy, environmental quality, and public health. This includes working to develop artificial intelligence decision-making algorithms for smart cities and working, in collaboration with social scientists, to engage citizens in the development and refinement of these algorithms.
- Develop and evaluate innovative approaches to reducing indoor and outdoor air pollution.

GRAND CHALLENGE 5:

Foster Informed Decisions and Actions

Addressing the world's largest environmental problems will require major shifts in our approaches and actions.[259] New strategies and technologies will only be effective in solving these grand challenges with widespread adoption, which may require regulatory changes at the governmental level and behavioral changes at the community and individual levels. For this to happen, decision makers in the public and private sectors and a significant portion of the general public must believe that the environmental problems are serious enough to warrant change—and that proposed solutions are worth adopting. In other words, addressing grand environmental challenges requires, in addition to effective solutions, a pervasive recognition that implementing those solutions is in our best interest.

Achieving this will require, first, engendering a civil society that is well informed about how the environment affects human well-being and prosperity. This is not about changing people's preferences or making the public "care" more about the environment. Rather, it is about equipping people with options that provide solutions and with information to make wise choices based on an understanding of the potential outcomes and costs associated with each course of action and the potential risks from inaction.

Second, it is important that experts and stakeholders act in partnership to identify problems and consider alternative solutions. There is sometimes a gap between what scientists and engineers believe will be useful for stakeholders and what the stakeholders themselves understand as useful.[260] It is possible to reduce this gap by taking a collaborative approach that engages both experts and stakeholders to define and prioritize problems, select alternatives to be considered, identify constraints and criteria for success, and consider issues of equity and distribution.

These first two elements—understanding and stakeholder engagement—create a foundation for identifying and implementing policy, management, and regulatory approaches to promote outcomes that are consistent with the collective priorities. Although the responsibility for engaging stakeholders and fostering full understanding of environmental choices does not lie entirely with environmental engineers, there is much that the engineering community can contribute.

Understanding Linkages Between the Environment, Human Well-Being, and Prosperity

In the context of environmental challenges, understanding potential consequences involves making the connection between our actions (or inactions) and the impacts that these have on the environment and the well-being of different individuals or groups in society. The choices made by individuals or groups can have spillover impacts on the well-being of others. For example, consider a property developer in an urban area deciding on the design of a new building and surrounding landscape. Incorporating features such as green or reflective roofs, reflective pavements, and increased tree plantings can reduce a property's contribution to urban heat island effects, but doing so often comes at a cost to the developer.[261]  Similarly, farmers deciding how much nitrogen fertilizer to apply will typically consider the benefit from improved yields and the cost of purchasing and applying the fertilizer. But applying nitrogen fertilizer has additional costs, such as when fertilizer leaches into surface water or groundwater, polluting a nearby town's water supply or downstream estuaries.[262] Some of the excess nitrogen will volatilize in the form of nitrous oxide, a powerful greenhouse gas, or to ammonia and nitrogen oxides, potentially contributing to regional air pollution.[263] The developers or farmers may be unaware of the impacts of their choices on the well-being of others. But even if they are aware, they typically have inadequate incentives to reduce environmental impacts because many of the consequences are borne by others (what economists refer to as "externalities").

Identifying and quantifying the full set of consequences of human actions on the environment and human well-being are active areas of research involving environmental engineering, ecology and other natural sciences, ecological and environmental economics, and other social sciences. Uncovering the important impacts often involves active cogeneration of knowledge by stakeholders and experts, as discussed in more depth in the next section. Over the past two decades, ecologists working with many other disciplines have made substantial progress in describing the benefits that nature provides to people, known as ecosystem services.[264] Ecosystem services include:

BOX 5-1. KAMEHAMEHA SCHOOLS: ANALYZING ECOSYSTEM SERVICES AND ENGAGING STAKEHOLDERS TO IMPROVE LAND-USE DECISIONS

In Hawaii, the largest private landowner is the education trust Kamehameha Schools, which owns roughly 8 percent of the land in the state. In the early 2000s, Kamehameha Schools faced a decision about what to do with a large block of land on the north shore of Oahu. Kamehameha Schools, engaged the Natural Capital Project[266] to analyze the effects of alternative land-use plans on carbon sequestration, water quality, and economic returns (see figure below). These endpoints and alternative land-use plans were developed in consultation with Kamehameha Schools and the local community with goals of balancing economic, environmental, educational, cultural, and community returns. A diversified agriculture land use was ultimately selected as the plan that best met the overall goals, even though monetized income returns were the lowest for this scenario. Kamehameha Schools was awarded the American Planning Association's 2011 National Planning Excellence Award for Innovation in Sustaining Places.[267]

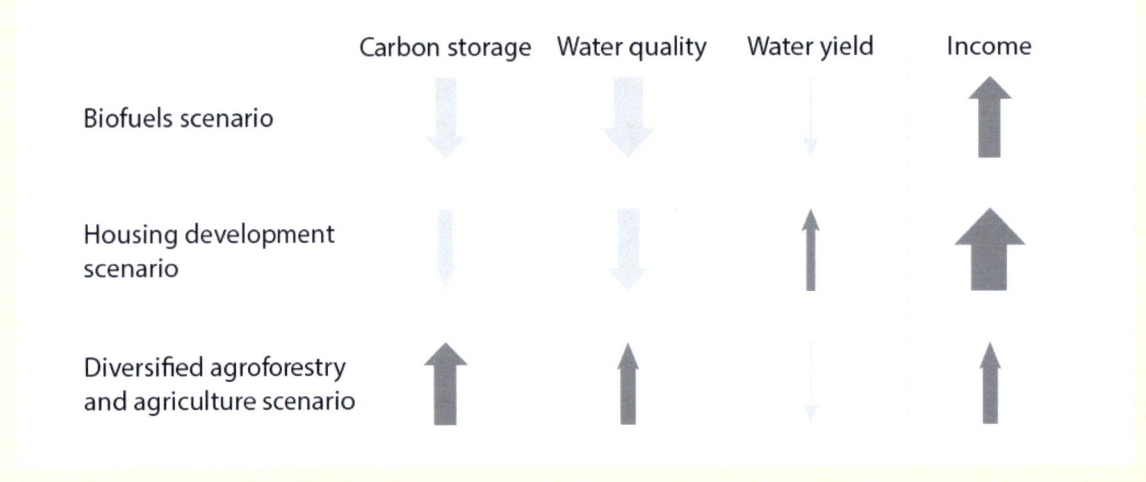

Projected changes in ecosystem services under three future land-use scenarios.

1. The provision of material goods (food, fiber, energy, and other materials);
2. Ecosystem functions that naturally regulate environmental conditions in ways that improve human living conditions, such as filtering pollutants from water or air, providing protection for coastal communities from storm surge, or reducing riverine flooding; and
3. Nonmaterial services related to psychological, spiritual, and cultural values.

Work on ecosystem services has highlighted many ways in which environmental protection or improvement can provide tangible benefits for human quality of life and prosperity (see Box 5-1). In addition, such work can highlight the risks posed by continued environmental degradation, including the potential for crossing thresholds with sudden catastrophic changes that may be difficult or impossible to reverse.[265] In the face of such risks, increasing system resilience is an important component of system design.

Another approach to quantifying the full set of environmental consequences is life-cycle assessment. This technique, commonly used by industrial ecologists and environmental engineers, aims to measure environmental impacts associated with producing and consuming specific products, from production of raw materials to the disposal of the product at the end of its useful life.[268] Life-cycle assessments

often measure impacts in physical units, such as materials and energy consumed or the amount of carbon dioxide emitted, and do not require assessment of impacts in monetary terms. This simplifies the analysis in some respects but can make it difficult to compare alternatives that have different types of environmental impacts. Other tools are also available to quantify the full environmental consequences of actions and to help engage stakeholders in this process (Box 5-2).

BOX 5-2. TOOLS TO CLARIFY SOCIAL, ENVIRONMENTAL, AND ECONOMIC DIMENSIONS OF CHOICES

A number of tools are used to help decision makers measure, monetize, or evaluate the potential impacts of a decision or action, including multiple social, environmental, and economic dimensions. Some tools help identify a full range of consequences of a given action. In addition to *life-cycle analyses, social impact assessments* identify possible social effects of an intervention or action.

Complex decisions often come down to weighing benefits against costs or risks and perhaps most importantly who pays the costs and who reaps the benefits (including intergenerational considerations). Tools to help clarify such decisions include *risk assessments* and *economic benefit-cost analyses*. *Chemical-alternatives assessment* evaluates hazards to human health and the environment of comparable chemicals (functionally) to choose the safest alternative. *Environmental-justice analysis* evaluates exposure and risk for minority populations and low-income populations to inform equitable decision making.

A number of stakeholder engagement tools are being used to facilitate collaboration and to ensure that multiple viewpoints are considered. *Collaborative problem solving* brings together stakeholders to work on a particular concern that has been identified. *Design charrettes* help stakeholders develop a mutually agreed-on vision of future development, usually regarding land-use planning decisions.

More than one tool can be applied simultaneously. The U.S. Environmental Protection Agency's Design for the Environment program[272] uses a variety of tools as it screens new chemicals, including collaborative problem solving with manufacturers and chemical alternatives assessments. The use of collaborative problem solving in conjunction with environmental-justice analysis helped officials in northeast Ohio make decisions on the best infrastructure options to help meet stormwater discharge limits and to provide additional environmental and recreational benefits using green infrastructure, particularly in low-income communities.[273]

The field of environmental economics has devoted decades of research to assessing the benefits of environmental improvement.[269] To make it easier to compare alternatives, economists typically try to measure all benefits and associated costs of environmental improvement in monetary terms, using market and nonmarket valuation techniques. For example, even though there is no market price for clean air, economists infer the value of clean air to homeowners by observing how home

values vary with air quality while controlling for other characteristics of houses that influence value, such as the lot size and number of bedrooms. However, some environmental impacts are difficult to measure in monetary terms, such as a community's sense of place or the value of the existence of other species. In addition, this approach can require a great deal of time and resources, and improved methods are needed to appropriately apply estimates developed in one area to other related areas.[270]

Because of the difficulty of quantifying all benefits in monetary terms, some business and environmental groups have pushed for a "triple bottom line" approach that captures environmental impacts, social responsibility, and financial returns without forcing all aspects to be evaluated in monetary terms.[271] Ideally, these assessments include metrics that reflect various values that are easily understood by stakeholders, such as measures of health impacts, water and air quality, biodiversity, and resilience. Using the triple bottom line approach to choose among alternative management or policy options would typically require a decision maker to weigh the relative importance of the three bottom lines.

Despite substantial progress toward understanding and quantifying the various impacts of our actions on the environment, important questions remain. For example:

- How do changes in policy and technology shape behavior in ways that affect the environment?
- How can knowledge from natural sciences, social sciences, and engineering disciplines be better integrated to understand how environmental change affects human well-being and prosperity?
- How can well-being and prosperity be measured in a rigorous and consistent manner and reported in a way that is readily understood by decision makers and stakeholders?

In addition, there is a great need to improve data collection to support robust ecosystem service analyses, life-cycle assessments, and other environmental analyses. This work should include consideration of the differential impacts on vulnerable communities and geographies due to physical, social, and economic factors. A significant part of this challenge is learning how to communicate clearly with decision makers and the broader community about the findings of environmental assessments and how various stakeholders value different benefits and costs.

Engaging With Stakeholders to Create Solutions

If progress is to be made toward the grand challenges, society must develop the right solutions for the right problems. The grand challenges identified in this report will manifest differently in different communities, and many efforts to address these challenges will play out on a local scale. Different communities have different values and priorities, and these should inform how problems are identified and addressed. In addition, the solutions that work in one community may not work in another. For successful adoption, it is crucial that innovations and approaches be acceptable and usable by the communities for which they are intended.

This cannot be achieved by scientists and engineers working in a vacuum. When considering specific strategies, multidisciplinary teams need to determine the circumstances under which they are most likely to be implemented, both in the near future and under a variety of future scenarios. What are the barriers to adoption and the potential for misuse? What are the economic, environmental, and social impacts of implementing these new strategies, including possible unintended consequences? How will the benefits and costs be distributed among different groups?

Research has shown that solutions are more likely to be successfully developed and adopted when interested stakeholders are engaged in a genuine dialogue with scientists and engineers that allows for iteration and exchange between the producers and users of research and technology (Figure 5-1). Such a process helps to better define the problem to be addressed, improves the likelihood that the priorities of various stakeholder groups are understood, and ensures that a broad range of alternatives are considered. Engaging with stakeholders can often reveal social or institutional factors that may affect the long-term success or failure of a new technology or strategy. It also reduces misunderstanding, increases perceived credibility, and generates trust.[274]

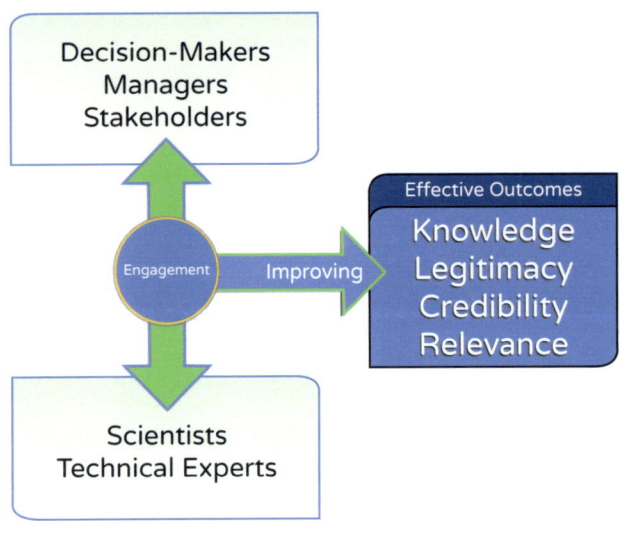

Figure 5-1. Effective public engagement on complex environmental challenges requires technical experts to learn from stakeholders and decision makers through a genuine two-way dialogue.

Distrust of science and technology are deeply held in many communities due to complex social, economic, and political forces, and this can present major barriers to the development of sustainable solutions. Like other members of the scientific community, engineers should work to understand the historical and political contexts behind these perspectives and identify opportunities for establishing new partnerships between engineers and stakeholders. Many well-intentioned scientists and engineers have focused their efforts on improving scientific understanding with the expectation that this will overcome skepticism. Yet, decades of social science research suggest that scientific literacy and technical knowledge have relatively minor impacts on the public's trust in science.[275] Rather than a lack of knowledge, skepticism more often stems from doubts regarding the honesty and integrity of outside experts and the institutions that they represent or concerns over the implications of proposed actions to their economic interests.

To overcome these tensions, engineers, scientists, and other experts should collaborate to forge relationships within skeptical communities, especially with trusted community leaders, to identify acceptable pathways forward. Transparency and inclusiveness should be prioritized in all aspects of the process, from data collection to decision making, by creating genuine opportunities for public participation, especially within communities that are seemingly disinterested, disadvantaged, or marginalized.[276]

Engineers should also strive to improve gender, racial, and ethnic diversity within the engineering community. Currently, African Americans, Hispanic Americans, and Native Americans are underrepresented in environmental engineering, and no gains have been made in increasing the percentage of undergraduate or graduate degrees awarded to underrepresented minorities in environmental engineering since 2008.[277] A community of engineers that represents and reflects the heterogeneous cultural and demographic backgrounds of society at large is necessary to understand the perspectives and interests of a diverse public. These varying life experiences will lead to the development of innovative strategies and technologies that may not necessarily come from a homogeneous group with similar world views.[278] In addition, improving professional opportunities for those from underrepresented backgrounds will bring in new talent and perspectives from wider segments of the population, generate healthy competition, and foster creativity.

Adopting Policy Solutions

The development of policy, both public and private, can encourage society to act with a full understanding of environmental impacts and long-term community priorities. Without policy interventions that help align private incentives to match societal objectives, the decisions and behaviors of individuals and businesses often do not account for externalities that are imposed on others. Most policy, management, and regulatory approaches relevant to addressing environmental challenges involve one or more of four basic elements: providing information, changing the decision context, creating incentives, and setting rules and regulations. Often a mix of these approaches is optimal.[279] In all four areas, research from the social and behavioral sciences combined with environmental engineering and science can help craft policies that are grounded in evidence and most likely to change behavior. Determining the best policy solutions to complex challenges, such as adapting to climate change, also often requires decision making under uncertainty, as discussed in Challenge 2.

Providing Information

Educating the public can be an effective strategy to drive widespread action or attitude change.[280] Successful public information campaigns, such as those launched to raise awareness of the problems associated with smoking or forest fires, clearly state the problem and provide simple actions that can be taken to address the problem ("only you can prevent forest fires"). Information can also be used to create social pressures that encourage change. For example, electricity bills that present a household's energy consumption relative to their neighbors have successfully reduced energy demand in many communities.[281]

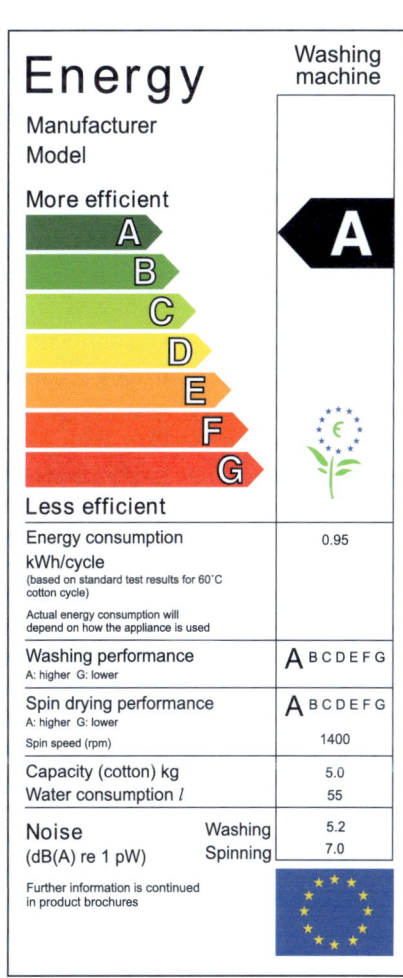

Figure 5-2. Manufacturer labeling is one strategy used to increase public awareness of environmental impacts and inform consumer choices.

In the context of complex environmental problems, information-based policies can take many forms including mandating disclosure, identifying chemicals of concern,

and advocating for transparency throughout the supply chain. For example, the "Ecolabel" program, developed in the European Union, identifies products that meet established environmental criteria considering a product's full life cycle. Governments, manufacturers, and retailers can make the environmental impacts of various products more transparent by supporting labels and collecting, curating, and sharing data (Figure 5-2). Environmental awareness could be further increased through calculation and expanded disclosure of carbon, water, and chemical footprints, supported by consensus-based standards and third-party audits. These efforts can in turn encourage innovation throughout the supply chain.

Subtle changes to the way information is presented can also have profound impacts by reducing known biases in decision making. For example, consumers systematically misunderstand fuel efficiency information when it is shown in miles per gallon. When the same information is depicted in gallons per mile, consumers make better financial and environmental choices.[282]

Changing the Decision Context

Policies that focus on changing behavior by modifying the decision context have gained much attention in recent years.[283] Such strategies are often designed to make a desired behavior easier or more probable by removing barriers to behavioral change. For example, reducing the paperwork and hassle required for a homeowner to participate in an energy efficiency rebate program can dramatically increase participation.[284] The percentage of people who have agreed to donate their organs is more than 95 percent higher in Belgium than in neighboring Netherlands, largely because citizens of Belgium are asked to sign a form to *opt out* of donating an organ, whereas citizens of the Netherlands must sign a form to *opt-in*.[285] Even though most citizens in both countries support organ donation, the hassle of opting in or opting out of a program can result in dramatic societal-level impacts on health and well-being. Careful consideration of default settings has resulted in improved environmental, health, and financial outcomes.[286]

Reducing barriers to behavior change is often less expensive and more politically feasible than other alternatives.[287] A challenge in implementing this approach, however, is for policy makers, social scientists, and engineers to collaboratively identify where such opportunities lie.

Creating Incentives

Policies can also be used to provide incentives for environmental solutions with broad societal benefits or disincentives for activities that contribute to environmental problems (see sidebar). Economic incentives are particularly valuable if technologies that provide broad ecosystem services come at a higher cost than similar technologies that do not. For example, tax credits have been provided to consumers to incentivize the purchase of electric cars and solar panels and to companies investing in renewable energy sources. Further, the government can take steps to reduce policy and financial risks for environmentally beneficial projects, such as by issuing partial loan guarantees and streamlining the permitting process, to make them more competitive with conventional projects among private investors.[288] Establishing disincentives for actions that are harmful to the environment is also an important policy mechanism. For example, if wetland impacts cannot be avoided as part of a permitted construction project, the Clean Water Act requires that other wetland areas be established or restored as compensation. Levees on carbon emissions could be established to discourage carbon emissions and stabilize the changing climate, while also funding permanent carbon sequestration efforts.

Social science research is needed to better understand how people respond to incentives. For example, are social incentives more or less effective than financial incentives for particular objectives? How can incentives or disincentives be implemented effectively and efficiently, considering their monitoring and enforcement costs? Environmental engineering research can inform policy makers about the systemwide benefits and costs of various policy alternatives.

INCENTIVIZING WATER CONSERVATION WITH SMART SOLAR PUMPS

Growth in the use of solar-powered pumps has reduced the energy costs of pumping water to irrigate crops while reducing carbon emissions. However, heavily subsidized solar panels and free solar power led to another problem: excessive irrigation of crops and overuse of limited groundwater supplies. Faced with this problem, researchers from the International Water Management Institute developed a solution that is partly technological and partly based on policy and management. Their "smart solar pump" initiative, piloted in Gujarat, India, incentivizes farmers to sell excess solar power back to the grid. Through guaranteed solar buy-back, farmers supplement their income, the country expands its energy reserves while making strides toward its renewable energy goals, and groundwater resources are conserved.[289]

Solar pump in Jagadhri, India.

Foster Informed Decisions and Actions | 75

Setting Rules and Regulations

Local, state, national, or international rules and regulations represent another tool to discourage environmental impacts and encourage improvements. For example, the Montreal Protocol in 1987 led to an international ban on the use of chlorofluorocarbons, which had damaged Earth's protective stratospheric ozone layer, and the result is that the ozone hole is now healing. Several countries have implemented bans on phosphate in detergents to address phosphorus pollution in surface waters. In the United States, after more than half of the states adopted phosphate detergent bans, the industry voluntarily removed phosphate from laundry detergents.[290] Another policy approach is to set environmental performance standards for government or corporate contracting and purchasing decisions, which would provide incentives for alternative technology choices and further technology development. Environmental rules and regulations are built upon substantial scientific and engineering research, and these efforts benefit from clear communication of policy-relevant scientific findings.

What Environmental Engineers Can Do

To foster informed decisions and actions, environmental engineers should work in collaboration with decision makers, stakeholders, and other experts to increase the public's understanding of the consequences of their choices, identify problems, create solutions, and support efforts to develop effective policies. Environmental engineers have the skills to assess the broad risks and benefits of alternative approaches to address the grand challenges and to work across disciplines as integrators of information. To develop approaches that are effective and acceptable—and therefore likely to succeed—it is vital to partner with

communities (particularly traditionally marginalized communities), businesses, and governments and work in collaboration with experts in social and behavioral sciences, communications, environmental and ecological economics, computer science, policy and management, and other disciplines. Given the complexity of the challenges to be addressed, it is to be expected that continuous iteration will be needed to refine collaborative approaches and develop feasible, acceptable, and impactful solutions.

To meet this challenge and create solutions that meet the needs of all, environmental engineers will need to build new skills and proactively diversify the field, as discussed in more detail in the next chapter. Examples of specific opportunities for environmental engineers to help address this challenge are highlighted in Box 5-3.

BOX 5-3. EXAMPLE AREAS FOR ENVIRONMENTAL ENGINEERS TO HELP FOSTER INFORMED DECISIONS AND ACTIONS

Some of the many ways environmental engineers can partner with other experts and stakeholders to help foster informed decisions and actions include:

- Work with communities and other disciplinary experts, including ecologists, economists, sociobehavioral scientists, and communication experts, to analyze and clearly communicate the potential consequences of alternative choices associated with the environmental grand challenges. Analyses should include impacts and benefits to individuals and various groups in society so that stakeholders and decision makers can better understand the impacts of their choices.
- Proactively diversify the field by recruiting members of underrepresented groups to become experts in the environmental engineering field and partner disciplines.
- Develop new approaches and technologies to collect environmental data needed to support ecosystem services and life-cycle analyses.
- Partner with communities and citizens to collect and assess environmental and socioeconomic data, understand the connections between trends and individual, corporate, and governmental behavior, and communicate the implications of this information. Environmental engineers can also develop enhanced participatory science approaches and technology-enabled platforms. Particular attention should be given to communities that have traditionally been underserved and marginalized.
- Develop transparent, user-friendly decision tools that can assist decision making by synthesizing information on financial, social, and environmental risks, costs, and benefits.

THE ULTIMATE CHALLENGE FOR ENVIRONMENTAL ENGINEERING:

Preparing the Field to Address a New Future

Historically, the discipline of environmental engineering has centered around public health and sanitation, and its practitioners' primary objectives have been to provide clean water and properly manage waste. These services are vital for the health and prosperity of society, lengthening life spans, and improving quality of life. The world now faces a number of challenges that are fundamentally broader in scope and larger in scale than the problems that environmental engineers have solved in the past. Communities have grown larger than ever. Technological innovation and major social changes occur over the course of years, rather than decades. Humans now influence the environment on a global as well as a local scale.

Environmental engineers should respond to the grand challenges outlined in this report and provide leadership to address them. To do so, the environmental engineering field must expand its scope, moving from a focus on individual problems toward systems-based solutions that address a broad set of issues. Environmental engineers will need to anticipate problems rather than react to them. The knowledge, skills, and problem-solving approaches that environmental engineers used in the past are not fully sufficient to meet the demands of the future. To create solutions that work for society, environmental engineers will need to cultivate diversity and engage collaboratively with stakeholders and other disciplines.

Some of these shifts are already under way. Environmental engineers are applying their expertise in areas beyond the field's roots in sanitation—air quality, green manufacturing, climate change, and urban design are examples (Figure 6-1). Environmental engineers have begun to evolve from those who characterize, manage, and remediate existing environmental problems to those who develop new knowledge, design innovative technologies and strategies, and implement solutions that prevent environmental problems. As this journey continues, environmental engineers can enable the creation of systems and infrastructure that allow people and ecosystems to thrive in the face of predictable and unforeseen challenges.

Adopting a new model for the discipline and practice of environmental engineering does not mean abandoning a proud history or eschewing traditional expertise. Environmental engineering can build on its strengths while positioning itself to keep pace with the scope and scale of society's needs. The following sections outline ways that environmental engineering practice, education, and research will need to evolve to best serve communities and address the complex global challenges ahead.

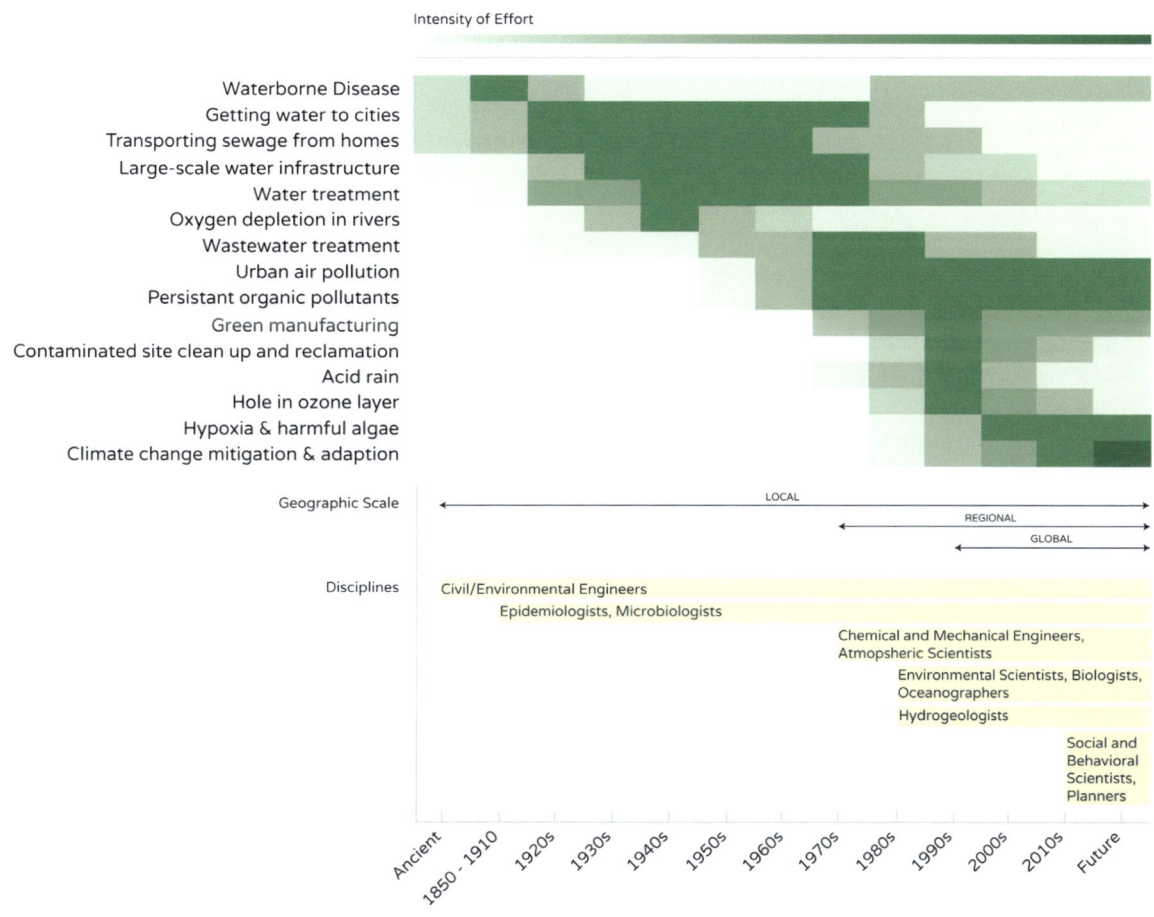

FIGURE 6-1. Timeline of major U.S. environmental engineering efforts, highlighting the broadening scale and complexity of the challenges and the expanding numbers of disciplines involved.

Environmental Engineering Practice

Environmental engineers operate in many sectors—private practice, industry, government, academia, and nonprofits—and at many scales, from individual facilities to international organizations. Environmental engineers apply their craft to a wide range of areas and develop careers in a variety of employment structures. How can environmental engineers evolve practices across this diverse field to address the complex challenges we face? While specific needs may vary for different sectors and areas of focus, there are two common threads. First, life-cycle and systems thinking should be applied within all aspects of environmental engineering to design or analyze solutions, considering the broad potential environmental, societal, and financial outcomes. Second, genuine engagement with communities and stakeholders and interdisciplinary collaboration with other experts are needed to inform solutions that are effective and likely to be implemented.

An example that encompasses both of these elements is the decision-making process used to address pathogens in the New York City water supply, as discussed in Challenge 4. A narrow view by traditional water treatment practice would be to focus solely on improving engineered water filtration. Instead, decision makers took a broader view. After evaluating multiple approaches to prevent or remove pathogens, they ultimately concluded that partnering with upstream residents and farmers to protect the watershed and compensating them for their efforts, along with more modest treatment, would be the most cost-effective approach to meet the water quality objectives while providing additional watershed benefits. Developing and implementing this approach required effective engagement of all stakeholders and ultimately reduced costs.

Solving global challenges cannot be done by environmental engineers alone. Today's environmental engineers work within complex, interconnected systems, often in the context of competing demands from agriculture, industry, people, and ecosystems. Environmental engineers need to work collaboratively with interdisciplinary partners and engage the public in the development of solutions. In most cases, there is no single answer that works for all communities, and solutions may need to be adapted over time. Environmental engineers need to examine the challenges and the alternative solutions using community input and considering short- and long-term consequences across local, regional, and global scales. In this role, the profession can provide a broad systems perspective for the disciplines and communities that are building the future so that they can more effectively achieve success. An engineering profession that represents the diversity of society at large will ensure that varied perspectives are understood and that the field draws upon the full potential of talent available.

> To effectively address the changes ahead, environmental engineering practitioners should work collaboratively with stakeholders and other disciplines to analyze, design, and implement practical, systems-based solutions. To support these efforts, the environmental engineering field should cultivate a more diverse workforce, focusing especially on increasing the racial and ethnic diversity of the pipeline.

Examples of steps that environmental engineers can take to transition to more collaborative and systems-based practices include the following:

1. Enhance stakeholder engagement by seeking diverse sources of information and community input.
2. Make use of available tools to incorporate full-cost life-cycle analysis and other sustainability tools to help stakeholders and decision makers understand the potential consequences of decision alternatives.
3. Build new tools to understand and predict adaptive and emergent behaviors of complex systems.
4. Implement evidence-based tools to recruit underrepresented minorities and women into the field beginning at the K-12 level and extending through graduate training.

Environmental Engineering Education

Although there are multiple models for educating and training environmental engineers, the 4-year undergraduate engineering program has traditionally served as the foundation for environmental engineering and is typically the minimum education required for practice. Most undergraduate environmental engineering programs today initially emphasize fundamental knowledge in the basic and applied sciences, mathematics, and engineering. This foundation is usually followed by more advanced courses focused on fate, transport, and treatment of contaminants in air, land, or water, and additional topics such as environmental health or the impacts of pollutants on ecosystems. Finally, through advanced courses or capstone projects, students explore subspecialties of the field and hone the skills that they will ultimately use in professional practice.[291] Many of these undergraduate students continue their environmental engineering education in graduate school to receive additional training and specialization. Because of the complex multidisciplinary nature of the field, environmental engineering has among the highest percentage of engineering practitioners with graduate degrees (50 percent based on 2014 data).[292]

Meeting tomorrow's more complex, integrated, global, and nuanced challenges will call for additional knowledge and skills beyond what today's environmental engineering curriculum provides. Educational institutions need to work with thought leaders from academic and practitioner communities and beyond to strengthen the education of tomorrow's environmental engineers, enhancing the curriculum and building essential skills.

Enhancing the Curriculum

To address society's environmental challenges, environmental engineers will need to have strength in their area of expertise but also have sufficient breadth to appreciate the broader societal context and devise effective solutions (Figure 6-2). For example, they will need to appreciate the social and behavioral components of the challenges they are trying to solve; even an efficient and effective technological

FIGURE 6-2. A T-shaped environmental engineer brings engineering depth with breadth in topics such as social science and policy that are essential to understanding and developing effective solutions for today's complex challenges.

solution cannot realize its full potential without consideration of societal, cultural, economic, legal, and political issues (see Challenge 5). To anticipate potential outcomes and avoid unintended consequences, environmental engineers will also need to understand the nonlinear and dynamic forces in many natural and human systems and the feedbacks that these forces can create.

Students also need opportunities for in-depth education in the scientific subspecialties that are most relevant to the future challenges. While environmental engineering programs provide a robust foundation in water and contaminants (in keeping with the historic focus of the field), topics such as climate, air pollution, and energy are sparsely covered in most current university curricula, leading to knowledge gaps between what our education system provides and the challenges future environmental engineers will face. Furthermore, most current environmental engineering curricula lack sufficient training in data science, which is emerging as a key strategy in 21st century solutions.

> To complement and build on the traditional emphasis on applied science, mathematics, and engineering, environmental engineering education programs should strengthen foundational knowledge in two areas: complex system dynamics and the social and behavioral dimensions of environmental challenges. In addition, programs should ensure that the scientific content of their curricula keeps pace with current and anticipated global challenges and the most promising tools for developing solutions.

Examples of steps that can be taken to enhance the foundational content of environmental engineering programs include the following:

- Cultivate rigorous systems thinking by integrating training in complex systems, data science, and decision analysis into the environmental engineering curriculum.
- Engage with colleagues in the social sciences to develop learning opportunities relevant to understanding the social, cultural, economic, legal, policy, and political contexts of environmental engineering challenges.

- Strengthen scientific curricula and subspecialty offerings for topics relevant to the full spectrum of current and anticipated challenges, such as climate, energy, and air pollution, in addition to more traditional areas of focus.

Building Essential Skills

In addition to these new areas of foundational knowledge, tomorrow's environmental engineers will require new types of skills, capabilities, and perspectives. Finding solutions that are feasible within the broader context also requires engaging decision makers and the public, and working collaboratively with experts in other disciplines. To complement the traditional focus on problem-solving skills, environmental engineering programs should educate students to communicate well and work collaboratively in interdisciplinary multicultural teams. In practice, it is rare for a single engineering solution to engender unanimous support. Environmental engineers must therefore learn to think creatively and critically, balance competing needs and priorities, forge compromises, and communicate persuasively. These capabilities are enhanced by an understanding of how people and communities make decisions, especially in the context of uncertainty, as well as a sense of empathy and social conscience.

> Environmental engineering education should equip graduates with the skills to communicate effectively, work collaboratively, think critically, and forge compromises.

Examples of steps educators can take to equip trainees with these skills include the following:

- Teach communication skills such as analyzing a communication situation, assessing the communication capacity and needs of target audiences, establishing goals and objectives, and formulating strategies.
- Develop partnerships with practitioners and community leaders to develop student learning experiences that involve real-world projects that are solved with creativity, stakeholder engagement, consensus building, and compromise.
- Provide opportunities for aspiring environmental engineers to directly experience community decision-making processes.
- Incorporate culturally relevant and diverse approaches to educational experiences, including activities that challenge students to develop solutions specific to socioeconomically disadvantaged and underserved communities.
- Create opportunities to explore the ethical and social dimensions of environmental engineering challenges.
- Offer educational experiences in negotiation, compromise, and conflict resolution.

Approaches to Engineering Education Reform

Solving the grand challenges in environmental engineering demands a broader approach to education. Interdisciplinary, experiential learning equips students to consider how myriad factors such as budget constraints, historical context, public acceptance, and regulatory frameworks affect the design and implementation of technological solutions to societal challenges. A new model for

environmental engineering education is needed to support the development of more innovative, creative, and effective problem solvers.

There is already movement in this direction. Several universities have instituted engineering leadership initiatives and developed educational models that broaden the undergraduate engineering experience.[293] The National Academy of Engineering also has led several efforts to advance undergraduate engineering education through its work with the Engineer of 2020 Project[294] and the Grand Challenge Scholars Program.[295] These initiatives illustrate how existing engineering programs can be enhanced without necessarily adding new coursework. For example, the Grand Challenges Scholars program involves training on five basic components—research/creative, multidisciplinary, business/entrepreneurship, multicultural, and social consciousness—that can be layered upon or crafted within existing degree requirements. By encouraging universities to adopt project- or service-based learning models to provide students with experience designing solutions in the context of real-world complexities, the Grand Challenge Scholars Program engenders a broader educational experience without fundamentally disrupting existing programs. This type of program could extend to significant global experiences with the goal of creating more effective and engaged environmental engineers.

There are limits to how much can be included in a 4-year undergraduate program. New subspecialties related to the grand challenges may need to be introduced at the undergraduate level but fully delivered through graduate programs. Engineering education can also be enhanced through other opportunities for formal and informal education. Continuing education that develops specialized knowledge and skills for practicing engineers is particularly important. Extracurricular activities involving experiential learning, such as student projects, study abroad, internships, independent research, student professional societies, and community involvement programs would benefit undergraduate and graduate programs.

Engineering programs are outliers in professional education in that they conflate the necessary preparation for practice with a general college education. Other

professions have abandoned this approach, transitioning their degree requirements to graduate programs to accommodate the needs of an increasingly demanding and specialized profession while enabling a comprehensive pre-professional college education. Physical therapy, pharmacy, and nursing are examples.[296] To cultivate a professional workforce with the breadth and depth necessary to address grand challenges, greater emphasis on graduate education in environmental engineering may be needed. By shifting specialization out of the undergraduate experience, this approach would also open opportunities for those with an undergraduate degree in engineering to pursue further education in other fields, such as law, policy, education, and economics, thus bringing a rich engineering background to those fields and enhancing opportunities for interdisciplinary collaboration. However, such an approach could have an undesired result of further reducing the percentage of underrepresented minorities in the field. If greater emphasis is placed on graduate programs to educate the environmental engineers of the future, targeted programs may be needed to recruit underrepresented groups into environmental engineering graduate programs.

> College and university programs should evaluate their undergraduate and graduate degree requirements and other educational opportunities to ensure that environmental engineers can receive sufficient training to address the grand challenges. Several approaches are available to broaden and strengthen the undergraduate environmental engineering education, but these changes may necessitate a greater reliance on graduate education for specialization.

Among the engineering disciplines, environmental engineering is well positioned to lead the charge toward a broader, more holistic approach to education. Examples of steps that educational institutions can take include:

- Restructure programs with greater reliance on graduate school for specialized training to allow undergraduates more room to explore topics such as the social and behavioral dimensions of environmental challenges, complex systems dynamics, data science, and real-world problem solving and to build the skills needed to develop solutions to complex interdisciplinary challenges. Such a change would likely require concurrent efforts to recruit and retain underrepresented minorities.
- Use practice- or service-based learning models to encourage experiential learning and enhance the undergraduate experience.
- Incorporate the Grand Challenge Scholars Program into undergraduate education.

Environmental Engineering Research

Research will continue to play a central role in propelling the technological innovations and the approaches needed to address society's grand challenges. Environmental engineering research is carried out in a variety of settings. Universities are perhaps the most visible contributors (and the primary backdrop for the analysis and vision outlined in this section), but national laboratories, government agencies, private corporations, nonprofit organizations, and international organizations and collaborations are also vital hubs for research and innovation.

The purpose of research in engineering is to increase the body of knowledge and discover better ways to solve problems. Underlying these overarching goals, there are two key factors that influence what types of research questions are pursued,

how research is carried out, and how the results are translated into practice. The first driver relates to employment structures for researchers. The second relates to research funding. While employment and funding structures vary across sectors, generally speaking, most researchers start by acquiring specialized foundational knowledge and research experience through formal education, obtain a research position within their subdiscipline, and then advance their career by independently spearheading projects, securing research funding, and publishing findings. In the United States, research funding, by and large, flows from the federal government into universities and other research organizations; however, many private companies also perform research.

Although these structures have produced significant gains, there is a growing awareness of ways in which the research enterprise falls short,[297] thus limiting engineering research—and consequently, engineering practice—from reaching its full potential. One of the most significant and pressing challenges in both research and practice is the need for effective interdisciplinary collaboration. Solving the grand challenges of the future will require advances within traditional environmental engineering disciplines but also engagement across the engineering disciplines, natural sciences, social sciences, and humanities. Environmental engineers today routinely collaborate with other engineering and science disciplines, but genuine collaboration with the social sciences is essential to developing effective solutions to 21st century challenges (see Challenge 5). Interdisciplinary research integrates information, perspectives, and techniques from multiple disciplines to address problems that cannot be fully addressed within a single discipline.[298]

Incentivizing Interdisciplinary Research

Successful interdisciplinary collaboration requires a cultural transition to embrace new areas of expertise and new ways of thinking, reinforced by incentives that provide tangible rewards for interdisciplinary work from the scale of the individual to the scale of the institution. This transition has been under way for approximately two decades,[299] but barriers remain.[300] Many university early career scholars are counseled to avoid interdisciplinary and team-based research based on the rationale that researchers should demonstrate strength within their discipline before engaging outside their discipline. This is reinforced by employment structures (particularly in universities) in which research positions are allocated to departments organized around a traditional discipline; researchers earn recognition and promotions by prioritizing independent scholarship in that discipline. Winning sole-investigator research grants and publishing with their students (rather than colleagues) in disciplinary journals is a far surer way to earn tenure and promotions than participating in large collaborations, for which papers take a long time to publish and have author lists comprising numerous investigators. These factors combine to make interdisciplinary research a liability for some early career scholars, despite the recent growth in funding, job opportunities, and, most importantly, the potential for substantial impact from such collaborations.

> Despite recent progress, when evaluating accomplishments many universities continue to prioritize disciplinary endeavors at the expense of interdisciplinary collaboration. To facilitate the collaboration necessary to meet future challenges, research employment structures should evolve to value and incentivize interdisciplinary work.

Examples of actions that research institutions can take to incentivize interdisciplinary collaboration include the following:

- Develop recruiting, promotion, and reward processes that reflect the interdisciplinary nature of environmental engineering, including valuing impact associated with coauthorship and publication in nontraditional interdisciplinary journals.[301]
- Enhance interdisciplinary mentoring to support the development and impact of early career scholars in nontraditional academic units and careers.

Interdisciplinary Research Support

Research support is a key factor for building and sustaining research programs and developing the next generation of scholars. Some of the primary agencies supporting research in the United States, including the National Science Foundation (NSF), still rely on disciplinary structures to organize their research agendas. However, crosscutting funding support for interdisciplinary scholarship has improved in recent years, stimulating a new generation of innovative research. Examples include NSF programs in Smart and Connected Communities, Innovations in Food, Energy and Water Systems, Dynamics of Coupled Natural and Human Systems, and Water Sustainability and Climate—programs that overlap substantially with the challenges presented in this report. Research support for early career scholars, such as the NSF Faculty Early Career Development Program (CAREER) and the Graduate Research Fellowship Program, remain primarily discipline based. Opportunities such as these can dramatically shape the long-term trajectory of a scholar.

Research teams supported by interdisciplinary initiatives often include engineers, social scientists, economists, and other experts. Success of these teams is contingent on genuine integration across disciplines. This requires more than placing experts from different disciplines into a room and adding interdisciplinary

verbiage to a proposal. Addressing today's interdisciplinary challenges requires investigators to invest time to build connections with others outside of their discipline, and universities need to value and reward this investment of time and effort. Although seed funding can incentivize new collaborations and reduce the barriers to launching new interdisciplinary projects, some of the most successful interdisciplinary collaborations are those that develop organically from groups of researchers inspired to address a common problem. For example, forums that bring interdisciplinary scholars together to discuss and present research can catalyze discussion and launch new collaborations. Though effective, these types of interactions are not yet commonplace. Interdisciplinary institutes can be well-suited for engaging diverse scholars around a common theme, provided that the institute itself does not become a silo.

> Interdisciplinary collaborations require meaningful interactions and genuine integration across disciplines. Funding organizations and research institutions can facilitate effective collaboration through well-designed grant programs and by fostering environments where relationships and collaborations can develop organically.

Examples of steps that research agencies, organizations, and corporations can take to foster interdisciplinary collaboration include the following:

- Craft opportunities for research support for early career scholars geared toward crosscutting and interdisciplinary themes.
- Prioritize expansion of interdisciplinary research support, even at the expense of disciplinary support, and incorporate proposal evaluation techniques that reward research teams and proposals that ensure genuine collaboration among scholars.[302]
- Develop NSF Engineering Research Centers focused around grand challenges, as recommended in the 2017 National Academies report, *A New Vision for Center-Based Engineering Research*.[303]

- Support workshops and other forums that stimulate interdisciplinary engagement and discussion around the grand challenges in environmental engineering.
- Embrace interdisciplinary research structures and programs that bring together researchers with different disciplinary expertise but common interests around specific challenges or themes.[304]

Industry and Community Engagement

The impact of interdisciplinary research is realized when new knowledge created by interdisciplinary teams is put into practice in industry and communities. Prototype applications at field scale provide opportunities to validate and refine new knowledge gained through scholarly research, and they provide a gateway for translation for societal benefit. These interactions also provide faculty with opportunities to understand the practical and fundamental issues that challenge engineering practice. Many university scholars have limited experience in translating research knowledge to applications at field scale. University promotion and reward programs often do not acknowledge the significance of these efforts, despite their importance. Several programs provide opportunities for both basic and translational research engaging academics and industrial partners in teams that solve real-world problems, including the Clinical and Translational Science Awards Program at the National Institutes of Health and NSF's Grant Opportunities for Academic Liaisons with Industry and Partnerships for Innovation programs. Examples of steps that research agencies, organizations, and corporations can take to enhance the translation of research to practice include the following:

- Develop additional opportunities for translating interdisciplinary research into practice through collaborative partnerships between industry, academia, and communities.
- Develop promotion and reward systems at universities to incentivize faculty engagement in translating environmental engineering research findings into practice, with emphasis on research products and technologies that have a demonstrated positive impact on society.

ENDNOTES

1. Cutler, D., and G. Miller. 2005. The role of public health improvements in health advances: The 20th century United States. *Demography* 42(1): 1-22.
2. U.S. Environmental Protection Agency. 2018. Air Pollutant Emissions Trends Data.
3. Engelhaupt, E. 2008. Happy birthday, Love Canal. *Chemical & Engineering News* 86(46): 46-53.
4. U.S. Environmental Protection Agency. 1992. CERCLA/Superfund Orientation Manual. EPA/542/R-92/005.
5. GBD 2016 Mortality Collaborators. 2017. Global, regional, and national under-5 mortality, adult mortality, age-specific mortality, and life expectancy, 1970–2016: A systematic analysis for the Global Burden of Disease Study 2016. *The Lancet* 390(10100): 1084-1150; Kontis, V., J. E. Bennett, C. D. Mathers, G. Li, K. Foreman, and M. Ezzati, 2017. Future life expectancy in 35 industrialised countries: Projections with a Bayesian model ensemble. *The Lancet* 389(10076): 1323-1335.
6. United Nations Department of Economic and Social Affairs. 2017. World Population Projected to Reach 9.8 Billion in 2050, and 11.2 Billion in 2100.
7. World Bank. 2017. No Poverty. Atlas of Sustainable Development Goals. Washington, DC.
8. World Health Organization and United Nations Children's Fund. 2017. *Progress on Drinking Water, Sanitation and Hygiene: 2017 Update and SGD Baselines*. Geneva: WHO and UNICEF.
9. International Energy Agency. 2017. *Energy Access Outlook 2017: From Poverty to Prosperity*. Organisation for Economic Co-operation and Development.
10. World Health Organization. 2018. Household Air Pollution and Health. Fact Sheet.
11. GBD 2016 Risk Factors Collaborators. 2017. Global, regional, and national comparative risk assessment of 84 behavioural, environmental and occupational, and metabolic risks or cluster of risks, 1990-2016: A systematic analysis for the Global Burden of Disease Study 2016. *The Lancet* 390(10100): 1345-1422.
12. United Nations Conference on Trade and Development. 2016. *Development and Globalization: Facts and Figures 2016*.
13. Kharas, H. 2010. The Emerging Middle Class in Developing Countries. OECD Development Centre Working Paper No. 285. Paris: Organisation for Economic Co-operation and Development.
14. United Nations. 2018. Sustainable Development Goals.
15. United Nations. 2017. *The Sustainable Development Goals Report 2017*.
16. World Health Organization. 2017. Children: Reducing Mortality. Fact Sheet. October.
17. World Health Organization and United Nations Children's Fund. 2017. *Progress on Drinking Water, Sanitation andHygiene: 2017 Update and SGD Baselines*. Geneva: WHO and UNICEF.
18. International Energy Agency. 2018. Access to Electricity. Energy Access Database.
19. World Health Organization and United Nations Children's Fund. 2017. *Progress on Drinking Water, Sanitation and Hygiene: 2017 Update and SGD Baselines*. Geneva: WHO and UNICEF.
20. Foley, J., N. Ramankutty, C. Balzer, E. M. Bennett, K. A. Brauman, S. R. Carpenter, E. Cassidy, J. Gerber, J. Hill, M. Johnston, C. Monfreda, N. D. Mueller, C. O'Connell, S. Polasky, D. K. Ray, J. Rockström, J. Sheehan, S. Siebert, D. Tilman, P. C. West, and D. P. M. Zaks. 2011. Solutions for a cultivated planet. *Nature* 478(7369): 337-342.
21. Mateo-Sagasta, J., S. M. Zadeh, and H. Turral. 2017. *Water Pollution from Agriculture: A Global Review*. Executive Summary. Rome: Food and Agriculture Organization of the United Nations, and Colombo, Sri Lanka: International Water Management Institute.
22. Food and Agriculture Organization of the United Nations. 2011. "Energy-Smart" Food for People and Climate. Issue Paper. Rome: FAO.
23. Godfray, H., J.R. Beddington, I.R. Crute, L. Haddad, D. Lawrence, J.F. Muir, J. Pretty, S. Robinson, S.M. Thomas, and C. Toulmin. 2010. Food security: The challenge of feeding 9 billion people. *Science* 327:812–818; Conway, G. 2012. One Billion Hungry: Can We Feed the World? Ithaca, NY: Cornell University.
24. Food and Agriculture Organization of the United Nations. 2017. *The Future of Food and Agriculture—Trends and Challenges*. Rome: FAO; U.S. Global Change Research Program. 2017. *Climate Science Special Report: Fourth National Climate Assessment*, Vol. 1. D. J. Wuebbles, D. W. Fahey, K. A. Hibbard, D. J. Dokken, B. C. Stewart, and T. K. Maycock, eds. Washington, DC: USGCRP.
25. National Academies of Sciences, Engineering, and Medicine. 2016. *Attribution of Extreme Weather Events in the Context of Climate Change*. Washington, DC: The National Academies Press; Reddy, V. R., S. K. Singh, and V. Anbumozhi. 2016. Food Supply Chain Disruption Due to Natural Disasters: Entities, Risks, and Strategies for Resilience. Economic Research Institute for ASEAN and East Asia Discussion Paper Series.
26. Foley, J. A., N. Ramankutty, K. A. Brauman, E. S. Cassidy, J. S. Gerber, M. Johnston, N. D. Mueller, C. O'Connell, D. K. Ray, P. C. West, and C. Balzer. 2011. Solutions for a cultivated planet. *Nature* 478(7369): 337-342.
27. Mahlein, A.-K. 2016. Plant disease detection by imaging sensors—Parallels and specific demands for precision agriculture and plant phenotyping. *Plant Disease* 100(2): 241-251.
28. Schumann, A. W. 2010. Precise placement and variable rate fertilizer application technologies for horticultural crops. *HortTechnology* 20(1): 34-40.
29. National Academies of Sciences, Engineering, and Medicine. 2018. *Science Breakthroughs to Advance Food and Agricultural Research by 2030*. Washington, DC: The National Academies Press.
30. National Academies of Sciences, Engineering, and Medicine. 2016. *Genetically Engineered Crops: Experiences and Prospects*. Washington, DC: The National Academies Press.
31. National Academies of Sciences, Engineering, and Medicine. 2018. *Science Breakthroughs to Advance Food and Agricultural Research by 2030*. Washington, DC: The National Academies Press.
32. Anand, G. 2016. Farmers' unchecked crop burning fuels India's air pollution. *The New York Times*. Nov. 2.
33. National Academies of Sciences, Engineering, and Medicine. 2018. *Scientific Breakthroughs to Advance Food and Agricultural Research by 2030*. Washington, DC: The National Academies Press.
34. Poppick, L. 2018. The future of fish farming may be indoors. *Scientific American*, September 17.
35. FAO. 2011. Global Food Losses and Food Waste—Extent, Causes, and Prevention. Rome: FAO.
36. Sharma, C., R. Dhiman, N. Rokana, and H. Panwar. 2017. Nanotechnology: An untapped resource for food packaging. *Frontiers in Microbiology* 12(8): 1735.
37. Gerber, P. J., H. Steinfeld, B. Henderson, A. Mottet, C. Opio, J. Dijkman, A. Falcucci, and G. Tempio. 2013. *Tackling Climate Change Through Livestock—A Global Assessment of Emissions and Mitigation Opportunities*. Rome: Food and Agriculture Organization.
38. Organisation for Economic Co-operation and Development and Food and Agriculture Organization. 2017. Commodity

snapshots: Meat. Pp. 110-112 in *OECD-FAO Agricultural Outlook 2017-2026. Special Focus: Southeast Asia.* Paris: OECD.
39. Ranganathan, J., D. Vennard, R. Waite, P. Dumas, B. Lipinski, and T. Searchinger. 2016. *Shifting Diets for a Sustainable Food Future.* Washington, DC: World Resources Institute.
40. Organisation for Economic Co-operation and Development. 2012. *OECD Environmental Outlook to 2050: The Consequences of Inaction.*
41. Gleick, P. H. 1993. Water in Crisis: *A Guide to the World's Fresh Water Resources.* New York: Oxford University Press.
42. van Dijk, A. I. J. M., H. E. Beck, R. S. Crosbie, R. A. M. de Jeu, Y. Y. Liu, G. M. Podger, B. Timbal, and N. R. Viney. 2013. The Millennium Drought in southeast Australia (2001–2009): Natural and human causes and implications for water resources, ecosystems, economy, and society. *Water Resources Research* 49(2): 1040-1057.
43. Swain, D. L., B. Langenbrunner, J. D. Neelin, and A. Hall. 2018. Increasing precipitation volatility in twenty-first-century California. *Nature Climate Change* 8(5): 427.
44. ScienceDaily. 2018. Water scarcity. Available at https://www.sciencedaily.com/terms/water_scarcity.htm.
45. Graf, W. 1999. Dam nation: A geographic census of American dams and their large-scale hydrologic impacts. *Water Resources Research* 5(4): 1305-1311; Richter, B. D., S. Postel, C. Revenga, T. Scudder, B. Lehner, A. Churchill, and M. Chow. 2010. Lost in development's shadow: The downstream human consequences of dams. *Water Alternatives* 3(2): 14-42; Gleick, P. H. 2000. A look at twenty-first century water resources development. *Water International* 25(1): 127-138.
46. Konikow, L. F. 2011. Contribution of global groundwater depletion since 1900 to sea-level rise. *Geophysical Research Letters* 38(17): L17401, doi: 10.1029/2011GL048604.
47. International Desalination Association. 2018. Desalination by the Numbers. Available at http://idadesal.org/desalination-101/desalination-by-the-numbers.
48. Dongare, P. D., A. Alabastri, S. Pedersen, K. R. Zodrow, N. J. Hogan, O. Neumann, J. Wu, T. Wang, A. Deshmukh, M. Elimelech, and Q. Li. 2017. Nanophotonics-enabled solar membrane distillation for off-grid water purification. *Proceedings of the National Academy of Sciences* 114(27): 6936-6941.
49. National Research Council. 2008. *Desalination: A National Perspective.* Washington, DC: The National Academies Press.
50. National Academies of Sciences, Engineering, and Medicine. 2016. *Using Graywater and Stormwater to Enhance Local Water Supplies: An Assessment of Risks, Costs, and Benefits.* Washington, DC: The National Academies Press.
51. National Research Council. 2012. *Water Reuse: Potential for Expanding the Nation's Water Supply Through Reuse of Municipal Wastewater.* Washington, DC: The National Academies Press.
52. National Research Council. 2012. *Water Reuse: Potential for Expanding the Nation's Water Supply Through Reuse of Municipal Wastewater.* Washington, DC: The National Academies Press.
53. National Academies of Sciences, Engineering, and Medicine. 2016. *Using Graywater and Stormwater to Enhance Local Water Supplies: An Assessment of Risks, Costs, and Benefits.* Washington, DC: The National Academies Press.
54. Maupin, M. A., J. F. Kenny, S. S. Hutson, J. K. Lovelace, N. L. Barber, and K. S. Linsey. 2014. Estimated Use of Water in the United States in 2010. U.S. Geological Survey Circular 1405.
55. U.S. Department of Agriculture, Economic Research Service. Ag and Food Statistics: Charting the Essentials. Available at https://www.ers.usda.gov/data-products/ag-and-food-statistics-charting-the-essentials/agricultural-trade/; Mekonnen, M. M., and A. Y. Hoekstra. 2011. The green, blue and grey water footprint of crops and derived crop products. *Hydrology and Earth System Sciences* 15(5): 1577.
56. Brauman, K. A., S. Siebert, and J. A. Foley. 2013. Improvements in crop water productivity increase water sustainability and food security—A global analysis. *Environmental Research Letters* 8(2): 024030.
57. Chaves, M. M., T. P. Santos, C. R. Souza, M. F. Ortuño, M. L. Rodrigues, C. M. Lopes, J. P. Maroco, J. S. Pereira. 2007. Deficit irrigation in grapevine improves water-use efficiency while controlling vigour and production quality. *Annals of Applied Biology* 150(2): 237-252.
58. Ali, M. H., and M. S. U. Taluder. 2008. Increasing water productivity in crop production—A synthesis. *Agricultural Water Management* 95(11): 1201-1213; National Academies of Sciences, Engineering, and Medicine. 2018. *Science Breakthroughs to Advance Food and Agricultural Research by 2030.* Washington, DC: The National Academies Press.
59. Food and Agriculture Organization of the United Nations. 2003. *Unlocking the Water Potential of Agriculture.*
60. Cutler, D., and G. Miller. 2005. The role of public health improvements in health advances: The 20th century United States. *Demography* 42(1): 1-22.
61. National Research Council. 2006. *Drinking Water Distribution Systems: Assessing and Reducing Risks.* Washington, DC: The National Academies Press.
62. Centers for Disease Control and Prevention. 2018. Legionella (Legionnaires' Disease and Pontiac Fever). Available at https://www.cdc.gov/legionella/about/history.html.
63. Pieper, K.J., L.A.H. Krometis, D.L. Gallagher, B.L. Benham, and M. Edwards. 2015. Incidence of waterborne lead in private drinking water systems in Virginia. *Journal of Water and Health* 13(3): 897-908; Deshommes, E., L. Laroche, S. Nour, C. Cartier, and M. Prévost. 2010. Source and occurrence of particulate lead in tap water. *Water Research* 44(12): 3734–3744.
64. United Nations. 2017. The Sustainable Development Goals Report 2017.
65. Bill & Melinda Gates Foundation. 2018. Reinvent the Toilet Challenge, Strategy Overview.
66. Gençer, E., C. Miskin. X. Sun, M. R. Khan, P. Bermel, M. A. Alam, and R. Agrawal. 2017. Directing solar photons to sustainably meet food, energy, and water needs. *Scientific Reports* 7:3133.
67. United Nations. 2017. The Sustainable Development Goals Report 2017.
68. U.S. Energy Information Administration. 2017. EIA projects 28% increase in world energy use by 2040. Today in Energy.
69. International Energy Agency. 2018. *The Future of Cooling: Opportunities for Energy Efficient Air Conditioning.* Organisation for Economic Co-operation and Development.
70. U.S. Energy Information Administration. 2018. Petroleum, natural gas, and coal still dominate U.S. energy consumption. Today in Energy.
71. International Energy Agency. 2017. Key World Energy Statistics. Available at https://www.iea.org/publications/freepublications/publication/KeyWorld2017.pdf.
72. National Research Council. 2010. *Hidden Costs of Energy: Unpriced Consequences of Energy Production and Use.* Washington, DC: The National Academies Press.
73. Jenner, S., and A. J. Lamadrid. 2012. Shale gas vs. coal: Policy implications from environmental impact comparisons of shale gas, conventional gas, and coal on air, water, and land in the United States. *Energy Policy* 53:442-453; U.S. Environmental Protection Agency. 2016. *Hydraulic Fracturing for Oil and Gas: Impacts from the Hydraulic Fracturing Water Cycle on Drinking Water Resources in the United States.* Final Report. EPA/600/R-16/236F. Washington, DC.
74. National Academies of Sciences, Engineering, and Medicine. 2017. *Flowback and Produced Waters: Opportunities and Challenges for Innovation: Proceedings of a Workshop.* Washington, DC: The National Academies Press.

75. National Academies of Sciences, Engineering, and Medicine. 2017. *Safely Transporting Hazardous Liquids and Gases in a Changing U.S. Energy Landscape*. Washington, DC: The National Academies Press.
76. National Research Council. 2007. *Environmental Impacts of Wind-Energy Projects*. Washington, DC: The National Academies Press; National Research Council. 2010. *Electricity from Renewable Resources: Status, Prospects, and Impediments*. Washington, DC: The National Academies Press.
77. National Research Council. 2007. *Environmental Impacts of Wind-Energy Projects*. Washington, DC: The National Academies Press.
78. American Wind Wildlife Institute. 2014. Wind Turbine Interactions with Wildlife and Their Habitats: A Summary of Research Results and Priority Questions. Fact Sheet.
79. National Research Council. 2010. *Hidden Costs of Energy: Unpriced Consequences of Energy Production and Use*. Washington, DC: The National Academies Press.
80. Sprecher, B., Y. Xiao, A. Walton, J. Speight, R. Harris, R. Kleijn, G. Visser, and G. J. Kramer. 2014. Life cycle inventory of the production of rare earths and the subsequent production of NdFeB rare earth permanent magnets. *Environmental Science & Technology* 48(7): 3951-3958.
81. Hill, J., E. Nelson, D. Tilman, S. Polasky, and D. Tiffany. 2006. Environmental, economic, and energetic costs and benefits of biodiesel and ethanol biofuels. *Proceedings of the National Academy of Sciences* 103(30): 11206-11210.
82. Lim, X. 2016. Uphill climb for biogas in Asia. *Chemical & Engineering News* 94:20-22.
83. Levin, T., and V.M. Thomas. 2016. Can developing countries leapfrog the centralized electrification paradigm? *Energy for Sustainable Development* 31: 97-107.
84. Koss, G. 2016. Renewable energy: Necessity drives Alaska's "petri dish" of innovation. *E&E News* Greenwire.
85. National Academies of Sciences, Engineering, and Medicine. 2017. *Enhancing the Resilience of the Nation's Electricity System*. Washington, DC: The National Academies Press.
86. Lawrence Berkeley National Laboratory. 2018. Microgrids at Berkeley Lab: Huatacondo. Available at https://building-microgrid.lbl.gov/huatacondo.
87. United National Conference on Trade and Development. 2017. The Least Developed Countries Report 2017: Transformational Energy Access. Geneva: UNCTAD/LDC/2017.
88. National Research Council. 2010. *The Power of Renewables: Opportunities and Challenges for China and the United States*. Washington, DC: The National Academies Press; National Research Council. 2010. *Electricity from Renewable Resources Status, Prospects, and Impediments*. Washington, DC: The National Academies Press.
89. Chen, H., Q. Ejaz, X. Gao, J. Huang, J. Morris, E. Monier, S. Paltsev, J. Reilly, A. Schlosser, J. Scott, and A. Sokolov. 2016. *Food, Water, Energy, Climate Outlook: Perspectives from 2016*. Massachusetts Institute of Technology Joint Program on the Science and Policy of Global Change.
90. National Academies of Sciences, Engineering, and Medicine. 2017. Enhancing the Resilience of the Nation's Electricity System. Washington, DC: The National Academies Press.
91. Penn, I. 2018. The $3 billion plan to turn Hoover Dam into a giant battery. *New York Times*, July 24.
92. Luo, X., J. Wang, M. Dooner, and J. Clarke. 2015. Overview of current development in electrical energy storage technologies and the application potential in power system operation. *Applied Energy* 137: 511-536.
93. Meadows, D. H. 2008. *Thinking in Systems: A Primer*. White River Junction, VT: Chelsea Green.
94. Mihelcic, J. R., J. B. Zimmerman, and M. T. Auer. 2014. *Environmental Engineering: Fundamentals, Sustainability, Design, Vol. 1*. Hoboken, NJ: Wiley.
95. Sterman. J. D. 1994. Learning in and about complex systems. System Dynamics Review 6(2-3): 291-330.
96. Boccara, N. 2010. Modeling Complex Systems, 2nd ed. New York: Springer.
97. National Academy of Sciences. 2014. *Climate Change: Evidence and Causes*. Washington, DC: The National Academies Press.
98. U.S. Global Change Research Program. 2017. *Climate Science Special Report: Fourth National Climate Assessment*, Vol. 1. D. J. Wuebbles, D. W. Fahey, K. A. Hibbard, D. J. Dokken, B. C. Stewart, and T. K. Maycock, eds. Washington, DC: USGCRP.
99. U.S. Global Change Research Program. 2017. *Climate Science Special Report: Fourth National Climate Assessment*, Vol. 1. D. J. Wuebbles, D. W. Fahey, K. A. Hibbard, D. J. Dokken, B. C. Stewart, and T. K. Maycock, eds. Washington, DC: USGCRP.
100. National Academies of Sciences, Engineering, and Medicine. 2017. *Attribution of Extreme Weather in the Context of Climate Change*. Washington, DC: The National Academies Press.
101. U.S. Global Change Research Program. 2017. *Climate Science Special Report: Fourth National Climate Assessment*, Vol. 1. D. J. Wuebbles, D. W. Fahey, K. A. Hibbard, D. J. Dokken, B. C. Stewart, and T. K. Maycock, eds. Washington, DC: USGCRP.
102. National Research Council. 2012. *Climate Change: Evidence, Impacts, and Choices*. Washington, DC: The National Academies Press.
103. Intergovernmental Panel on Climate Change. 2015. *Climate Change 2014: Mitigation of Climate Change. Contribution of Working Group III to the IPCC Fifth Assessment Report*. Cambridge, UK: Cambridge University Press.
104. U.S. Global Change Research Program. 2017. *Climate Science Special Report: Fourth National Climate Assessment*, Vol. 1. D. J. Wuebbles, D. W. Fahey, K. A. Hibbard, D. J. Dokken, B. C. Stewart, and T. K. Maycock, eds. Washington, DC: USGCRP.
105. Knoblauch C., C. Beer, S. Liebner, M. N. Grigoriev, and E. M. Pfeiffer. 2018. Methane production as key to the greenhouse gas budget of thawing permafrost. *Nature Climate Change* 8: 309-312.
106. Intergovernmental Panel on Climate Change. 2014. Climate Change 2014 Synthesis Report: Summary for Policymakers.
107. International Panel on Climate Change. 2018. Global warming of 1.5°C. *An IPCC special report on the impacts of global warming of 1.5°C above pre-industrial levels and related global greenhouse gas emission pathways, in the context of strengthening the global response to the threat of climate change, sustainable development, and efforts to eradicate poverty*. V. Masson-Delmotte, P. Zhai, H. O. Pörtner, D. Roberts, J. Skea, P. R. Shukla, A. Pirani, Y. Chen, S. Connors, M. Gomis, E. Lonnoy, J. B. R. Matthews, W. Moufouma-Okia, C. Péan, R. Pidcock, N. Reay, M. Tignor, T. Waterfield, and X. Zhou (eds.). In Press.
108. U.S. Global Change Research Program. 2017. *Climate Science Special Report: Fourth National Climate Assessment*, Vol. 1. D. J. Wuebbles, D. W. Fahey, K. A. Hibbard, D. J. Dokken, B. C. Stewart, and T. K. Maycock, eds. Washington, DC: USGCRP.
109. World Health Organization. 2018. Ambient (Outdoor) Air Quality and Health. Fact Sheet.
110. Williams, J. H., B. Haley, F. Kahrl, J. Moore, A. D. Jones, M. S. Torn, and H. McJeon. 2014. *Pathways to Deep Decarbonization in the United States*. [Revision with technical supplement. Nov 16, 2015].
111. Federal Ministry for Economic Affairs and Energy. 2016. Green Paper on Energy Efficiency. Berlin, Germany.
112. National Academies of Sciences, Engineering, and Medicine. 2010. *Real Prospects for Energy Efficiency in the United States*. Washington, DC: The National Academies Press.
113. U.S. Energy Information Administration. 2018. Electricity Explained: Electricity in the United States, Generation, Capacity, and Sales.

114. National Renewable Energy Laboratory. 2012. *Renewable Electricity Futures Study: Exploration of High-Penetration Renewable Electricity Futures*, Vol. 1. NREL/TP-6A20-52409. Golden, CO: NREL.
115. Cole, T. M., P. Donohoo-Vallett, J. Richards, and P. Das. 2017. *Standard Scenarios Report: A U.S. Electricity Sector Outlook*. NREL/TP-6A20-68548. Golden, CO: National Renewable Energy Laboratory.
116. International Panel on Climate Change. 2018. *Global warming of 1.5°C. An IPCC special report on the impacts of global warming of 1.5°C above pre-industrial levels and related global greenhouse gas emission pathways, in the context of strengthening the global response to the threat of climate change, sustainable development, and efforts to eradicate poverty*. V., Masson-Delmotte, P. Zhai, H. O. Pörtner, D. Roberts, J. Skea, P.R. Shukla, A. Pirani, Y. Chen, S. Connors, M. Gomis, E. Lonnoy, J. B. R. Matthews, W. Moufouma-Okia, C. Péan, R. Pidcock, N. Reay, M. Tignor, T. Waterfield, and X. Zhou (eds.)]. In Press.
117. International Energy Agency. 2017. *World Energy Outlook 2017*.
118. Rueter, G., and M. Kuebler. 2017. China leading the way in solar energy expansion as renewables surge. *Deutsche Welle*, July 6.
119. U.S. Department of Energy. Advanced Reacter Technologies https://www.energy.gov/ne/nuclear-reactor-technologies/advanced-reactor-technologies.
120. U.S. Energy Information Administration. 2018. *Use of Energy In the United States Explained: Energy Use for Transportation*. Available at https://www.eia.gov/energyexplained/?page=us_energy_transportation.
121. Zev Alliance. 2017. The rise of electric vehicles: The second million. Blog, Jan. 31. Available at http://www.zevalliance.org/second-million-electric-vehicles.
122. Lutsey, N., M. Grant, S. Wappelhorst, and H. Zhou. Power Play: How Governments Are Spurring the Electric Vehicle Industry. White Paper. Washington, DC: International Council on Clean Transportation.
123. Mucio, D. 2017. These countries are banning gas-powered vehicles by 2040. *Business Insider*, Oct. 23. Available at https://www.businessinsider.com/countries-banning-gas-cars-2017-10.
124. National Research Council. 2011. *Climate Stabilization Targets: Emissions, Concentrations, and Impacts over Decades to Millennia*. Washington, DC: The National Academies Press.
125. National Academies of Sciences, Engineering, and Medicine. 2018. *Negative Emissions Technologies and Reliable Sequestration: A Research Agenda*. Washington, DC: The National Academies Press.
126. National Academies of Sciences, Engineering, and Medicine. 2018. *Science Breakthroughs to Advance Food and Agricultural Research by 2030*. Washington, DC: The National Academies Press.
127. National Academies of Sciences, Engineering, and Medicine. 2018. *Negative Emissions Technologies and Reliable Sequestration: A Research Agenda*. Washington, DC: The National Academies Press.
128. Griscom, B. W., J. Adams, P. W. Ellis, R. A. Houghton, G. Lomax, D. A. Miteva, W. H. Schlesinger, D. Shoch, J. V. Siikamäki, P. Smith, P. Woodbury, C. Zganjar, A. Blackman, J. Campari, R. T. Conant, C. Delgado, P. Elias, T. Gopalakrishna, M. R. Hamsik, M. Herrero, J. Kiesecker, E. Landis, L. Laestadius, S. M. Leavitt, M. Minnemeyer, S. Polasky, P. Potapov, F. E. Putz, J. Sanderman, M. Silvius, E. Wollenberg, and J. Fargione. 2017. Natural climate solutions. *Proceedings of the National Academy of Sciences* 114(44): 11645-11650.
129. National Academies of Sciences, Engineering, and Medicine. 2018. Negative Emissions Technologies and Reliable Sequestration: A Research Agenda. Washington, DC: The National Academies Press.
130. Hunter, M. C., R. G. Smith, M. E. Schipanski, L.W. Atwood, and D.A. Mortensen. 2017. Agriculture in 2050: Recalibrating targets for sustainable intensification. *Bioscience* 67(4): 386-391.
131. National Research Council. 2015. *Climate Intervention: Reflecting Sunlight to Cool Earth*. Washington, DC: The National Academies Press.
132. U.S. Environmental Protection Agency. 2016. Greenhouse Gas Emissions: Overview of Greenhouse Gases.
133. U.S. Environmental Protection Agency. 2016. Global Methane Initiative: Importance of Methane.
134. National Academies of Sciences, Engineering, and Medicine. 2018. *Improving Characterization of Anthropogenic Methane Emissions in the United States*. Washington, DC: The National Academies Press.
135. Horowitz, J., and J. Gottlieb. 2010. The Role of Agriculture in Reducing Greenhouse Gas Emissions. Economic Brief No. 15. Washington, DC: U.S. Department of Agriculture Economic Research Service.
136. U.S. Global Change Research Program. 2017. *Climate Science Special Report: Fourth National Climate Assessment*, Vol. 1. D. J. Wuebbles, D. W. Fahey, K. A. Hibbard, D. J. Dokken, B. C. Stewart, and T. K. Maycock, eds. Washington, DC: USGCRP.
137. National Research Council. 2012. *Climate Change: Evidence, Impacts, and Choices*. Washington, DC: The National Academies Press.
138. Bates, B., Z. W. Kundzewicz, S. Wu, and J. Palutikof. 2008. Climate Change and Water. IPCC Technical Paper VI. Geneva: Intergovernmental Panel on Climate Change Secretariat.
139. National Academies of Sciences, Engineering, and Medicine. 2016. *Attribution of Extreme Weather Events in the Context of Climate Change*. Washington, DC: The National Academies Press.
140. Geophysical Fluid Dynamics Laboratory. 2018. Global Warming and Hurricanes: An Overview of Current Research Results. Princeton University Forrestal Campus.
141. Baltes, N.J., J. Gil-Humanes, and D.F. Voytas. 2017. Chapter One-Genome Engineering and Agriculture: Opportunities and Challenges. *Progress in Molecular Biology and Translational Science* 149: 1-26.
142. Phelan, P. E., K. Kaloush, M. Miner, J. Golden, B. Phelan, H. Silval II, and R. A. Taylor. 2015. Urban heat island: Mechanisms, implications, and possible remedies. *Annual Review of Environment and Resources* 40: 285-307.
143. Lempert, R. J., D. G., Groves, S. W., Popper, and S. C. Bankes. 2006. A General, Analytic Method for Generating Robust Strategies and Narrative Scenarios. *Management Science* 52(4): 514-528; Haasnoot, M., J. H. Kwakkel, W. E. Walker, and J. ter Maat. 2013. Dynamic adaptive policy pathways: A method for crafting robust decisions for a deeply uncertain world. *Global Environmental Change* 23(2): 485-498.
144. Westerling, A. L., B. P. Bryant, H. K. Preisler, T. P. Holmes, H. G. Hidalgo, T. Das, and S. R. Shrestha. 2011. Climate change and growth scenarios for California wildfire. *Climatic Change* 109(Supp. 1): 445-463; Barbero, R., J. T. Abatzoglou, N. K. Larkin, C. A. Kolden, and B. Stocks. 2015. Climate change presents increased potential for very large fires in the contiguous United States. *International Journal of Wildland Fire* 24(7): 892-899.
145. Smith, A., C. A. Kolden, T. B. Paveglio, M. A. Cochrane, D. M. Bowman, M. A. Moritz, A. D. Kliskey, L. Alessa, A. T. Hudak, C. M. Hoffman, J. A. Lutz, L. P. Queen, S. J. Goetz, P. E. Higuera, L. Boschetti, M. Flannigan, K. M. Yedinak, A. C. Watts, E. K. Strand, J. W. Van Wagtendonk, J. W. Anderson, B. J. Stocks, and J. T. Abatzoglou. 2016. The science of firescapes: Achieving fire-resilient communities. *BioScience* 66(2): 130-146.
146. National Research Council. 2012. *Disaster Resilience: A National Imperative*. Washington, DC: The National Academies Press.
147. Field, C. B., V. R. Barros, K. J. Mach, M. D. Mastrandrea, M. van Aalst, W. N. Adger, D. J. Arent, J. Barnett, R. Betts, T. E. Bilir, J. Birkmann, J. Carmin, D. D. Chadee, A. J. Challinor,

M. Chatterjee, W. Cramer, D. J. Davidson, Y. O. Estrada, J.-P. Gattuso, Y. Hijioka, O. Hoegh-Guldberg, H. Q. Huang, G. E. Insarov, R. N. Jones, R. S. Kovats, P. Romero-Lankao, J. N. Larsen, I.J. Losada, J. A. Marengo, R. F. McLean, L. O. Mearns, R. Mechler, J. F. Morton, I. Niang, T. Oki, J. M. Olwoch, M. Opondo, E. S. Poloczanska, H.-O. Pörtner, M. H. Redsteer, A. Reisinger, A. Revi, D. N. Schmidt, M. R. Shaw, W. Solecki, D. A. Stone, J. M. R. Stone, K. M. Strzepek, A. G. Suarez, P. Tschakert, R. Valentini, S. Vicuña, A. Villamizar, K. E. Vincent, R. Warren, L. L. White, T. J. Wilbanks, P. P. Wong, and G. W. Yohe. 2014. Technical summary. Pp. 35-94 in *Climate Change 2014: Impacts, Adaptation, and Vulnerability. Part A: Global and Sectoral Aspects. Contribution of Working Group II to the IPCC Fifth Assessment Report*. C. B. Field, V. R. Barros, D. J. Dokken, K. J. Mach, M. D. Mastrandrea, T. E. Bilir, M. Chatterjee, K. L. Ebi, Y. O. Estrada, R. C. Genova, B. Girma, E. S. Kissel, A. N. Levy, S. MacCracken, P. R. Mastrandrea, and L. L. White, eds. Cambridge, UK, and New York: Cambridge University Press.
148. Shiferaw, B., M. Smale, H. Braun, E. Duveiller, M. Reynolds, and G. Muricho. 2013. Crops that feed the world 10. Past successes and future challenges to the role played by wheat in global food security. *Food Security* 5(3): 291-317.
149. Howden, S. M., J. F. Soussana, F. N. Tubiello, N. Chhetri, M. Dunlop, and H. Meinke. 2007. Adapting agriculture to climate change. *Proceedings of the National Academy of Sciences* 104(50): 19691-19696; Smit, B., and M. W. Skinner. 2002. Adaptation options in agriculture to climate change: A typology. *Mitigation and Adaptation Strategies for Global Change* 7:85-114.
150. Field, C.B., V.R. Barros, K.J. Mach, M.D. Mastrandrea, M. van Aalst, W.N. Adger, D.J. Arent, J. Barnett, R. Betts, T.E. Bilir, J. Birkmann, J. Carmin, D.D. Chadee, A.J. Challinor, M. Chatterjee, W. Cramer, D.J. Davidson, Y.O. Estrada, J.-P. Gattuso, Y. Hijioka, O. Hoegh-Guldberg, H.-Q. Huang, G.E. Insarov, R.N. Jones, R.S. Kovats, P. Romero Lankao, J.N. Larsen, I.J. Losada, J.A. Marengo, R.F. McLean, L.O. Mearns, R. Mechler, J.F. Morton, I. Niang, T. Oki, J.M. Olwoch, M. Opondo, E.S. Poloczanska, H.-O. Pörtner, M.H. Redsteer, A. Reisinger, A. Revi, D.N. Schmidt, M.R. Shaw, W. Solecki, D.A. Stone, J.M.R. Stone, K.M. Strzepek, A.G. Suarez, P. Tschakert, R. Valentini, S. Vicuña, A. Villamizar, K.E. Vincent, R. Warren, L.L. White, T.J. Wilbanks, P.P. Wong, and G.W. Yohe. 2014. Technical Summary. *Climate Change 2014: Impacts, Adaptation, and Vulnerability. Part A: Global and Sectoral Aspects. Contribution of Working Group II to the Fifth Assessment Report of the Intergovernmental Panel on Climate Change* [Field, C.B., V.R. Barros, D.J. Dokken, K.J. Mach, M.D. Mastrandrea, T.E. Bilir, M. Chatterjee, K.L. Ebi, Y.O. Estrada, R.C. Genova, B. Girma, E.S. Kissel, A.N. Levy, S. MacCracken, P.R. Mastrandrea, and L.L. White (eds.)]. Cambridge University Press, Cambridge, United Kingdom and New York, NY, USA, pp. 35-94.
151. U.S. Global Change Research Program. 2017. *Climate Science Special Report: Fourth National Climate Assessment*, Vol. 1. D. J. Wuebbles, D. W. Fahey, K. A. Hibbard, D. J. Dokken, B. C. Stewart, and T. K. Maycock, eds. Washington, DC: USGCRP. doi: 10.7930/J0J964J6.
152. Milman, O. 2017. Atlantic City and Miami Beach: Two takes on tackling the rising waters. *The Guardian*, Mar. 20.
153. Katz, C. 2013. To Control Floods, The Dutch Turn to Nature for Inspiration. *Yale Environment 360*.
154. Coastal Protection and Restoration Authority of Louisiana. 2017. Louisiana's Comprehensive Master Plan for a Sustainable Coast. Coastal Protection and Restoration Authority of Louisiana. Baton Rouge, LA.
155. Watts, N., M. Amann, S. Ayeb-Karlsson, K. Belesova, T. Bouley, M. Boykoff, P. Byass, W. Cai, D. Campbell-Lendrum, J. Chambers, and P. M. Cox. 2017. The Lancet countdown on health and climate change: From 25 years of inaction to a global transformation for public health. *The Lancet* 391(10120): 581-630.
156. Haines, A. 2008. Climate change, extreme events, and human health. Pp. 57-74 in *Global Climate Change and Extreme Weather Events: Understanding the Contributions to Infectious Disease Emergence*. Washington, DC: The National Academies Press.
157. Zorrilla, C. D. 2017. The view from Puerto Rico—Hurricane Maria and its aftermath. *New England Journal of Medicine* 377(19): 1801-1803.
158. National Research Council. 2009. *Informing Decisions in a Changing Climate*. Washington, DC: The National Academies Press; Dittrich, R., A. Wreford, and D. Moran. 2016. A survey of decision-making approaches for climate change adaptation: Are robust methods the way forward? *Ecological Economics* 122: 79-89; Walker, W. E., M. Haasnoot, and J. H. Kwakkel. 2013. Adapt or perish: A review of planning approaches for adaptation under deep uncertainty. *Sustainability* 5(3): 955-979.
159. Matthews, E., C. Amann, S. Bringezu, W. Hüttler, C. Ottke, E. Rodenburg, D. Rogich, H. Schandl, E. Van, D. Voet, and H. Weisz. 2000. *The Weight of Nations: Material Outflows from Industrial Economies*. Washington, DC: World Resources Institute.
160. U.S. Environmental Protection Agency. 2018. National Overview: Facts and Figures on Materials, Wastes and Recycling. Trends—1960 to Today.
161. Ellen MacArthur Foundation. 2013. *Towards the Circular Economy: Economic and Business Rationale for an Accelerated Transition*, Vol. 1.
162. Hoornweg, D., and P. Bhada-Tata. 2012. *What a Waste: A Global Review of Solid Waste Management*. Washington, DC: World Bank.
163. Hoornweg, D., P. Bhada-Tata, and C. Kennedy. 2013. Environment: Waste production must peak this century. *Nature* 503: 615.
164. Kharas, H. 2017. The Unprecedented Expansion of the Global Middle Class: An Update. Global Economy and Development at Brookings. Working Paper 100. Brookings Institution.
165. National Research Council. 2005. *Contaminants in the Subsurface: Source Zone Assessment and Remediation*. Washington, DC: The National Academies Press.
166. National Research Council. 2005. *Contaminants in the Subsurface: Source Zone Assessment and Remediation*. Washington, DC: The National Academies Press.
167. National Research Council. 2013. *Alternatives for Managing the Nation's Complex Contaminated Groundwater Sites*. Washington, DC: The National Academies Press.
168. Health Effects Institute. 2015. The Advanced Collaborate Emissions Study (ACES). Executive Summary. Boston: HEI; Khalek, I. A., T. L. Bougher, P. M. Merritt, and B. Zielinska. 2011. Regulated and unregulated emissions from highway heavy-duty diesel engines complying with U.S. Environmental Protection Agency 2007 emissions standards. *Journal of the Air and Waste Management Association* 61(4): 427-442.
169. World Water Assessment Programme. 2009. *The United Nations World Water Development Report 3: Water in a Changing World*. Paris: UNESCO, and London: Earthscan, Table 8.1, p. 137.
170. International Food Policy Research Institute and VEOLIA. 2015. The Murky Future of Global Water Quality: New Global Study Projects Rapid Deterioration in Water Quality. White Paper. Washington, DC: IFPRI and Chicago: VEOLIA Water North America; World Health Organization. 2016. Air Pollution Levels Rising in Many of the World's Poorest Cities. News Release.
171. Walsh, J., D. Wuebbles, K. Hayhoe, J. Kossin, K. Kunkel, G. Stephens, P. Thorne, R. Vose, M. Wehner, J. Willis, D. Anderson, S. Doney, R. Feely, P. Hennon, V. Kharin, T. Knutson, F. Landerer, T. Lenton, J. Kennedy, and R. Somerville,

2014: Our changing climate. Pp. 19-67 in *Climate Change Impacts in the United States: The Third National Climate Assessment*, J. M. Melillo, T. C. Richmond, and G. W. Yohe, eds., U.S. Global Change Research Program. doi:10.7930/J0KW5CXT.
172. Lindstrom, A. B., M. J. Strynar, and E. L. Libelo. 2011. Polyfluorinated compounds: Past, present, and future. *Environmental Science & Technology* 45(19): 7954-7961.
173. Roser, M. 2018. Life Expectancy. Our World in Data. Available at https://ourworldindata.org/life-expectancy.
174. GBD 2016 Risk Factors Collaborators. 2017. Global, regional, and national comparative risk assessment of 84 behavioural, environmental and occupational, and metabolic risks or cluster of risks, 1990-2016: A systematic analysis for the Global Burden of Disease Study 2016. *The Lancet* 390(10100): 1345-1422.
175. GBD 2016 Risk Factors Collaborators. 2017. Global, regional, and national comparative risk assessment of 84 behavioural, environmental and occupational, and metabolic risks or cluster of risks, 1990-2016: A systematic analysis for the Global Burden of Disease Study 2016. *The Lancet* 390(10100): 1345-1422.
176. Health Effects Institute. 2018. State of Global Air. Available at: www.stateofglobalair.org.
177. Landrigan, P. J., R. Fuller, N. J. Acosta, O. Adeyi, R. Arnold, A. B. Baldé, R. Bertollini, S. Bose-O'Reilly, J. I. Boufford, P. N. Breysse, T. Chiles, C. Mahidol, A. M. Coll-Seck, M. L. Cropper, J. Fobil, V. Fuster, M. Greenstone, and M. Zhong. 2017. The Lancet Commission on pollution and health. *The Lancet* 391(10119): 462-512.
178. Cunningham, V. L., M. Buzby, T. Hutchinson, F. Mastrocco, N. Parke, and N. Roden. 2006. Effects of human pharmaceuticals on aquatic life: Next steps. *Environmental Science & Technology* 40(11):3456-3462; Iwanowicz, L. R., V. S. Blazer, A. E. Pinkney, C. P. Guy, A. M. Major, K. Munney, S. Mierzykowski, S. Lingenfelser, A. Secord, K. Patnode, T. J. Kubiak, C. Stern, C. M. Hahn, D. D. Iwanowicz, H. L. Walsh, and A. Sperry. 2016. Evidence of estrogenic endocrine disruption in smallmouth and largemouth bass inhabiting Northeast U.S. national wildlife refuge waters: A reconnaissance study. *Ecotoxicology and Environmental Safety* 124: 50-59.
179. Jambeck, J. R., R. Geyer, C. Wilcox, T. R. Siegler, M. Perryman, A. Andrady, R. Narayan, and K. L. Law. 2015. Plastic waste inputs from land into the ocean. *Science* 347(6223): 768-771.
180. Tosetto, L., C. Brown, and J. E. Williamson. 2016. Microplastics on beaches: Ingestion and behavioural consequences for beachhoppers. *Marine Biology* 163(10): 199; Nelms, S. E., T. S., Galloway, B. J. Godley, D. S. Jarvis, and P. K. Lindeque. 2018. Investigating microplastic trophic transfer in marine top predators. *Environmental Pollution* 238: 999-1007; World Economic Forum. 2016. *The New Plastics Economy: Rethinking the Future of Plastics*. Geneva, Switzerland.
181. Anderson, D. M., P. M. Glibert, and J. M. Burkholder. 2002. Harmful algal blooms and eutrophication: Nutrient sources, composition, and consequences. *Estuaries* 25(4): 704-726.
182. Michalak, A. M. 2016. Study role of climate change in extreme threats to water quality. *Nature* 535(7612): 349-352.
183. Mueller, R., and V. Yingling. 2017. History and Use of Per- and Polyfluoroalkyl Substances (PFAS). Fact Sheet. Interstate Technology Regulatory Council. November.
184. Mueller, R., and V. Yingling. 2018. Environmental Fate and Transport for Per- and Polyfluoroalkyl Substances. Fact Sheet. Interstate Technology Regulatory Council. March.
185. National Ground Water Association. 2018. PFAS: Top 10 Facts. Available at https://www.ngwa.org/docs/default-source/default-document-library/pfas/pfastop-10.pdf?sfvrsn=8c8ef98b_2.
186. Agency for Toxic Substances and Disease Registry. 2018. Toxicological Profile for Perfluoroalkyls: Draft for Public Comment, June. Available at https://www.atsdr.cdc.gov/toxprofiles/tp200.pdf.
187. China Council for International Cooperation on Environment and Development. 2014. Special Policy Study on Soil Pollution Management. Available at http://environmental-partnership.org/wp-content/uploads/2016/01/SPS-on-Soil-Pollution-Management.pdf.
188. Hu, Y., X. Liu, J. Bai, K. Shih, E. Y. Zeng, and H. Cheng. 2013. Assessing heavy metal pollution in the surface soils of a region that had undergone three decades of intense industrialization and urbanization. *Environmental Science and Pollution Research* 20(9): 6150-6159.
189. Hu, Y., H. Cheng, and S. Tao. 2016. The challenges and solutions for cadmium-contaminated rice in China: A critical review. *Environment International* 92–93: 515-532.
190. Coulon, F. K. Jones, H. Li, Q. Hu, J. Gao, F. Li, M. Chen, Y.-G. Zhu, R. Liu, M. Liu, K. Canning, N. Harries, P. Bardos, P. Nathanail, R. Sweeney, D. Middleton, M. Charnley, J. Randall, M. Richell, T. Howard, I. Martin, S. Spooner, J. Weeks, M. Cave, F. Yu, F. Zhang, Y. Jiang, P. Longhurst, G. Prpich, R. Bewley, J. Abra, and S. Pollard. 2016. China's soil and groundwater management challenges: Lessons from the UK's experience and opportunities for China. *Environment International*, 91: 196-200.
191. Song, Y., D. Hou, J. Zhang, D. O'Connor, G. Li, Q. Gu, S. Li, and P. Liu. 2018. Environmental and socio-economic sustainability appraisal of contaminated land remediation strategies: A case study at a mega-site in China. *Science of The Total Environment* 610-611: 391-401.
192. U.S. Environmental Protection Agency. 2000. Tetrachloroethylene (Perchloroethylene). Available at: https://www.epa.gov/sites/production/files/2016-09/documents/tetrachloroethylene.pdf.
193. http://zwia.org/.
194. U.S. Environmental Protection Agency. 1999. *Achieving Clean Air and Clean Water: Report of the Blue Ribbon Panel on Oxygenates in Gasoline*. EPA420-R-99-021.
195. Landrigan, P. J., R. Fuller, N. J. R. Acosta, O. Adeyi, R. Arnold, N. Basu, A. B. Baldé, R. Bertollini, S. Bose-O'Reilly, J. I. Boufford, P. N. Breysse, T. Chiles, C. Mahidol, A. M. Coll-Seck, M. L. Cropper, J. Fobil, V. Fuster, M. Greenstone, A. Haines, D. Hanrahan, D. Hunter, M. Khare, A. Krupnick, B. Lanphear, B. Lohani, K. Martin, K. V. Mathiasen, M. A. McTeer, C. J. L. Murray, J. D. Ndahimananjara, F. Perera, J. Potočnik, A. S. Preker, J. Ramesh, J. Rockström, C. Salinas, L. D. Samson, K. Sandilya, P. D. Sly, K. R. Smith, A. Steiner, R. B. Stewart, W. A. Suk, O. C. P. van Schayck, G. N. Yadama, K. Yumkella, and M. Zhong. 2018. The *Lancet* Commission on Pollution and Health. *The Lancet* 391(10119): 462-512.
196. Zeng, X., J. A. Mathews, and J. Li. 2018. Urban mining of e-waste is becoming more cost-effective than virgin mining. *Environmental Science & Technology* 52(8): 4835-4841; Nguyen, R. T., L. A. Diaz, D. D. Imholte, and T. E. Lister. 2017. Economic assessment for recycling critical metals from hard disk drives using a comprehensive recovery process. *JOM* 69(9): 1546-1552.
197. National Academies of Sciences, Engineering, and Medicine. 2018. *Gaseous Carbon Waste Streams Utilization: Status and Research Needs*. Washington, DC: The National Academies Press.
198. Deublein, D., and A. Steinhauser, eds. 2011. *Biogas from Waste and Renewable Resources: An Introduction*. Weinheim, Germany: Wiley VCH Verlag.
199. McCarty, P. L., J. Bae, and J. Kim. 2011. Domestic wastewater treatment as a net energy producer—Can this be achieved? *Environmental Science & Technology* 45(17): 7100-7106.
200. Water Environment Research Foundation. 2012. *Barriers to Biogas Use for Renewable Energy*. Report OWSO11C10. Alexandria, VA: WERF.

201. Smith, A. L., L. B. Stadler, L. Cao, N. G. Love, L. Raskin, and S. J. Skerlos. 2014. Navigating wastewater energy recovery strategies: A life cycle comparison of anaerobic membrane bioreactor and conventional treatment systems with anaerobic digestion. *Environmental Science & Technology* 48(10): 5972-5981.
202. Steffen, W., K. Richardson, J. Rockström, S. E. Cornell, I. Fetzer, E. M. Bennett, R. Biggs, S. R. Carpenter, W. De Vries, C. A. De Wit, C. Folke, D. Gerten, J. Heinke, G. M. Mace, L. M. Persson, V. Ramanathan, B. Reyers, and S. Sörlin. 2015. Planetary boundaries: Guiding human development on a changing planet. *Science* 347(6223): 1259855.
203. Jasinski, S. M. 2017. Phosphate rock. Mineral Commodity Summaries. U.S. Geological Survey.
204. Mihelcic, J. R., L. M. Fry, and R. Shaw. 2011. Global potential of phosphorus recovery from human urine and feces. *Chemosphere* 84(6): 832-839.
205. Larsen, T. A., A. C. Alder, R. I. L. Eggen, M. Maurer, and J. Lienert. 2009. Source separation: Will we see a paradigm shift in wastewater handling? *Environmental Science & Technology* 43(16): 6121-6125.
206. International Fertilizer Industry Association. 2009. *Energy Efficiency and CO_2 Emissions in Ammonia Production: 2008-2009 Summary Report*. Paris: IFIA.
207. National Academy of Engineering. 1997. *The Industrial Green Game: Implications for Environmental Design and Management*. Washington, DC: The National Academy Press.
208. U.S. Environmental Protection Agency. 2016. Advancing Sustainable Materials Management: 2014. Fact Sheet.
209. Organisation for Economic Co-operation and Development. 2015. *Environment at a Glance 2015: OECD Indicators*. Paris: OECD Publishing.
210. Baldé, C. P., V. Forti, V. Gray, R. Kuehr, and P. Stegmann. 2017. *The Global E-waste Monitor 2017: Quantities, Flows, and Resources*. Bonn, Geneva, and Vienna. United Nations University, International Telecommunication Union, and International Solid Waste Association.
211. U.S. Environmental Protection Agency. 2004. Evaluation Report: Multiple Actions Taken to Address Electronic Waste, but EPA Needs to Provide Clear National Direction. Office of the Inspector General, Report No. 2004-P-00028.
212. Hansen, T. L., J. la Cour Jansen, Å. Davidsson, and T. H. Christensen. 2007. Effects of pre-treatment technologies on quantity and quality of source-sorted municipal organic waste for biogas recovery. *Waste Management* 27(3): 398-405.
213. Lewis, J. J., and S. K. Pattanayak. 2012. Who adopts improved fuels and cookstoves? A systematic review. *Environmental Health Perspectives* 120(5): 637-645.
214. World Bank Group. 2017. Populations Estimates and Projections. Available at https://data.worldbank.org/data-catalog/population-projection-tables.
215. UN Habitat. 2016. *Urbanization and Development: Emerging Futures*. World Cities Report 2016. Nairobi, Kenya: United Nations Human Settlements Programme.
216. UN Habitat. 2016. *Urbanization and Development: Emerging Futures*. World Cities Report 2016. Nairobi, Kenya: United Nations Human Settlements Programme.
217. United Nations Environment Programme. 2012. Global Initiative for Resource Efficient Cities: Engine to Sustainability; UN Habitat. 2016. *Urbanization and Development: Emerging Futures*. World Cities Report 2016. Nairobi, Kenya: United Nations Human Settlements Programme.
218. Editorial. 2016. A missed opportunity for urban health. *The Lancet* 388(10056): 2057. doi:10.1016/S0140-6736(16)32056-6; Editorial. 2017. Health in slums: Understanding the unseen. *The Lancet* 389(10068): 478-479. doi:10.1016/S0140-6736(17)30266-0.
219. Ezeh, A., O. Oyebade, D. Satterthwaite, Y. F. Chen, R. Ndugwa, J. Sartori, B. Mberu, G. J. Melendez-Torres, T. Haregu, S. I. Watson, and W. Caiaffa. 2017. The history, geography, and sociology of slums and the health problems of people who live in slums. *The Lancet* 389(10068): 547-558; Landrigan, P. J., R. Fuller, N. J. Acosta, O. Adeyi, R. Arnold, A. B. Baldé, R. Bertollini, S. Bose-O'Reilly, J. I. Boufford, P. N. Breysse, and T. Chiles, C. Mahidol, A. M. Coll-Seck, M. L. Cropper, J. Fobil, V. Fuster, M. Greenstone, A. Haines, D. Hanrahan, D. Hunter, M. Khare, A. Krupnick, B. Lanphear, B. Lohani, K. Martin, K. V. Mathiasen, M. A. McTeer, C. J. L. Murray, J. D. Ndahimananjara, F. Perera, J. Potočnik, A. S. Preker, J. Ramesh, J. Rockström, C. Salinas, L. D. Samson, K. Sandilya, P. D. Sly, K. R. Smith, A. Steiner, R. B. Stewart, W. A. Suk, O. C. P. van Schayck, G. N. Yadama, K. Yumkella, and M. Zhong. 2018. The *Lancet* Commission on Pollution and Health. *The Lancet* 391(10119): 462-512.
220. Morse, S. S., J. A. K. Mazet, M. Woolhouse, C. R. Parrish, D. Carroll, W. B. Karesh, C. Zimbrana-Torrelio, W. I. Lipkin, and P. Daszak. 2012. Prediction and prevention of the next zoonosis. *The Lancet* 380(9857):1956-1965.
221. Revi, A., D. E. Satterthwaite, F. Aragón-Durand, J. Corfee-Morlot, R. B. R. Kiunsi, M. Pelling, D. C. Roberts, and W. Solecki. 2014. Urban areas. Pp. 535-612 in *Climate Change 2014: Impacts, Adaptation, and Vulnerability. Part A: Global and Sectoral Aspects. Contribution of Working Group II to the IPCC Fifth Assessment Report*. C. B. Field, V. R. Barros, D. J. Dokken, K. J. Mach, M. D. Mastrandrea, T. E. Bilir, M. Chatterjee, K. L. Ebi, Y. O. Estrada, R. C. Genova, B. Girma, E. S. Kissel, A. N. Levy, S. MacCracken, P. R. Mastrandrea, and L. L. White, eds. Cambridge, UK, and New York: Cambridge University Press.
222. American Society of Civil Engineers; Engineers. 2017. *2017 Infrastructure Report Card: A Comprehensive Assessment of American's Infrastructure*.
223. Organisation for Economic Co-operation and Development. 2007. *Infrastructure to 2030, Vol.2: Mapping Policy for Electricity, Water and Transport*. Paris: OECD Publishing.
224. National Academies of Sciences, Engineering, and Medicine. 2016. *Pathways to Urban Sustainability: Challenges and Opportunities for the United States*. Washington, DC: The National Academies Press; Ramaswami, A., A. Russell, P. Culligan, K. Sharma, and E. Kumar. 2016. Meta-principles for developing smart, sustainable, and healthy cities. *Science* 352(6288): 940-943.
225. Jeong, H., O. A. Broesicke, B. Drew, D. Li, and J. C. Crittenden. 2016. Life cycle assessment of low impact development technologies combined with conventional centralized water systems for the City of Atlanta, Georgia. *Environmental Science and Engineering* 10(6): 1-13.
226. New York State Department of Environmental Conservation. New York City Water Supply. Available at: www.dec.ny.gov/lands/25599.html.
227. Zanella, A., N. Bui, A. Castellani, L. Vangelista, and M. Zorzi. 2014. Internet of Things for smart cities. *IEEE Internet of Things Journal* 1(1): 22-32.
228. Debnath, A. K., H. C. Chin, M. M. Haque, and B. Yuen. 2014. A methodological framework for benchmarking smart transport cities. *Cities* 37: 47-56.
229. Ramaprasad, A., A. Sánchez-Ortiz, and T. Syn. 2017. A unified definition of a smart city. Pp. 13-24 in *Electronic Government*. M. Janssen, K. Axelsson, O. Glassey, B. Klievink, R. Krimmer, I. Lindgren, P. Parycek, H. J. Scholl, and D. Trutnev, eds. Springer, Cham.
230. World Economic Forum. 2018. *Harnessing Artificial Intelligence for the Earth*.
231. Palca, J. 2018. Betting on artificial intelligence to guide earthquake response. NPR, April 20. Available at: https://www.npr.org/

2018/04/20/595564470/betting-on-artificial-intelligence-to-guide-earthquake-response.

232. Laursen, L. 2014. Barcelona's smart city ecosystem. *MIT Technology Review*, Nov. 18.

233. Amsterdam Smart City. Smartphone app for citizens to manage street lighting. Available at: https://amsterdamsmartcity.com/products/amsterdam-offers-smartphone-app-for-cityzens-to-manage-street-lighting.

234. Korkali, M., J. G. Veneman, B. F. Tivnan, J. P. Bagrow, and P. D. Hines. 2017. Reducing cascading failure risk by increasing infrastructure network interdependence. *Scientific Reports* 7(44499).

235. Examples are Ecube Lab' (https://www.ecubelabs.com/solution); Bigbelly (http://bigbelly.com); and IBM. 2015. IBM Intelligent Waste Management Platform. White Paper. Available at: https://www-01.ibm.com/common/ssi/cgi-bin/ssialias?htmlfid=GVW03059USEN.

236. World Bank. 2015. How an open traffic platform is helping Asian cities mitigate congestion, pollution. News.

237. CrimeRadar. Frequently Asked Questions. Available at: https://rio.crimeradar.org/faq.

238. Sidewalk Labs. 2017. Vision Sections of RFP Submission. RFP No. 2017-13.

239. Woyke, E. 2018. A smarter smart city. *MIT Technology Review*, Feb. 21.

240. Sidewalk Labs. 2017. Vision Sections of RFP Submission. RFP No. 2017-13.

241. World Economic Forum. 2018. *Harnessing Artificial Intelligence for the Earth*.

242. National Academies of Sciences, Engineering, and Medicine. 2016. *Building Smart Communities for the Future: Proceedings of a Workshop—in Brief*. Washington, DC: The National Academies Press.

243. Klepeis, N. E., W. C. Nelson, W. R. Ott, J. P. Robinson, A. M. Tsang, P. Switzer, J. V. Behar, S. C. Hern, and W. H. Engelmann. 2001. The National Human Activity Pattern Survey (NHAPS): A resource for assessing exposure to environmental pollutants. *Journal of Exposure Science and Environmental Epidemiology* 11(3): 231-252.

244. Dai, D., A. J. Prussin II, L. C. Marr, P. J. Vikesland, M. A. Edwards, and A. Pruden. 2017. Factors shaping the human exposome in the built environment: Opportunities for engineering control. *Environmental Science & Technology* 51(14): 7759-7774.

245. Jones, K. E., N. G. Patel, M. A. Levy, A. Storeygard, D. Balk, J. L. Gittleman, and P. Daszak. 2008. Global trends in emerging infectious diseases. *Nature* 451(7181): 990-993.

246. Lerner, H., and C. Berg. 2017. A comparison of three holistic approaches to health: One Health, EcoHealth, and Planetary Health. *Frontiers in Veterinary Science* 4(163); Centers for Disease Control and Prevention. 2018. One Health Basics. Available at: https://www.cdc.gov/onehealth/basics.

247. Vikesland, P. J., A. Pruden, P. J. J. Alvarez, D. Aga, H. Burgmann, X. Li, C. M. Manaia, I. Nambi, K. Wigginton, T. Zhang, and Y. Zhu. 2017. Toward a comprehensive strategy to mitigate dissemination of environmental sources of antibiotic resistance. *Environmental Science & Technology* 51(22): 13061-13069.

248. Omira, A. 2016. Kibagare Haki Zetu Bio-Centre: A Transformation Story. Umande Trust, Aug. 8. Available at: http://umande.org/kibagare-haki-zetu-bio-centre-a-transformation-story.

249. P.L. 109-58; P.L. 111-364.

250. U.S. Environmental Protection Agency. 2017. Environmental Justice FY2017 Progress Report. 240-R1-8001.

251. Maintenance and Management Oversight Committee. Muddy River Restoration Project: Flood Control Improvement. Available at: http://www.muddyrivermmoc.org/flood-control.

252. C40 Cities. 2015. Cities100: Copenhagen—Creating a Climate Resilient Neighborhood. Available at: http://www.c40.org/case_studies/cities100-copenhagen-creating-a-climate-resilient-neighborhood.

253. Zhang, W., S. Guhathakurta, J. Fang, and G. Zhang. 2015. Exploring the impact of shared autonomous vehicles on urban parking demand: An agent-based simulation approach. *Sustainable Cities and Society* 19: 34-45.

254. UN Habitat. 2016. *Urbanization and Development: Emerging Futures*. World Cities Report 2016. Nairobi, Kenya: United Nations Human Settlements Programme.

255. Jeong, H., O. A. Broesicke, B. Drew, and J. C. Crittenden. 2018. Life cycle assessment of small-scale greywater reclamation systems combined with conventional centralized water systems for the City of Atlanta, Georgia. *Journal of Cleaner Production* 174: 333-342.

256. U.S. Environmental Protection Agency. 2015. Catalog of CHP Technologies. Available at: https://www.epa.gov/chp/catalog-chp-technologies.

257. James, J.-A., V. M. Thomas, A. Pandit, D. Li, and J. C. Crittenden. 2016. Water, air emissions, and cost impacts of air-cooled microturbines for combined cooling, heating, and power systems: A case study in the Atlanta region. *Engineering* 2(4):470-480; James, J.-A., S. Sung, H. Jeong, O. A. Broesicke, S. P. French, D. Li, and J. C. Crittenden. 2017. Impacts of combined cooling, heating, and power systems and rainwater harvesting on water demand, carbon dioxide and NOx emissions for Atlanta. *Environmental Science & Technology* 52:3-10.

258. MacKerron, G., and S. Mourato. 2013. Happiness is greater in natural environments. *Global Environmental Change* 23(5): 992-1000.

259. Guerry, A., S. Polasky, J. Lubchenco, R. Chaplin-Kramer, G. C. Daily, R. Griffin, M. H. Ruckelshaus, I. J. Bateman, A. Duraiappah, T. Elmqvist, M. W. Feldman, C. Folke, J. Hoekstra, P. Kareiva, B. Keeler, S. Li, E. McKenzie, Z. Ouyang, B. Reyers, T. Ricketts, J. Rockström, H. Tallis, and B. Vira. 2015. Natural capital informing decisions: From promise to practice. *Proceedings of the National Academy of Sciences* 112: 7348-7355.

260. Lemos, M. C., C. J. Kirchhoff, and V. Ramprasad. 2012. Narrowing the climate information usability gap. *Nature Climate Change* 2(11): 789-794.

261. Rizwan, A. M, D. Y. C. Leung, and C. Liu. 2008. A review on the generation, determination and mitigation of urban heat island. *Journal of Environmental Sciences* 20: 120-128; Phelan, P. E., K. Kaloush, M. Miner, J. Golden, B. Phelan, H. Silva III, and R. A. Taylor. 2015. Urban heat island: Mechanisms, implications, and possible remedies. *Annual Review of Environment and Resources* 40: 285-307.

262. Carpenter, S. R., N. F. Caraco, D. L. Correll, R. W. Howarth, A. N. Sharpley, and V. H. Smith. 1998. Nonpoint pollution of surface waters with phosphorus and nitrogen. *Ecological Applications* 8(3): 559-568.

263. Hill, J., S. Polasky, E. Nelson, D. Tilman, H. Huo, L. Ludwig, J. Neumann, H. Zheng, and D. Bonta. 2009. Climate change and health costs of air emissions from biofuels and gasoline. *Proceedings of the National Academy of Sciences* 106(6): 2077-2082.

264. National Research Council. 2005. *Valuing Ecosystem Services: Toward Better Environmental Decision-Making*. Washington, DC: The National Academies Press; Millennium Ecosystem Assessment. 2005. *Ecosystems and Human Well-Being: Synthesis*. Washington, DC: Island Press; Díaz, S., U. Pascual, M. Stenseke, B. Martín-López, R. T. Watson, Z. Molnár, R. Hill, K. M. A. Chan, I. A. Baste, K. A. Brauman, S. Polasky, A. Church, M. Lonsdale, A. Larigauderie, P. W. Leadley, A. P. E.

van Oudenhoven, F. van der Plaat, M. Schröter, S. Lavorel, Y. Aumeeruddy-Thomas, E. Bukvareva, K. Davies, S. Demissew, G. Erpul, P. Failler, C. A. Guerra, C. L. Hewitt, H. Keune, S. Lindley, and Y. Shirayama. 2018. An inclusive approach to assess nature's contributions to people. *Science* 359: 270-272.

265. Scheffer, M., S. R. Carpenter, J. A. Foley, C. Folke, and B. Walker. 2001. Catastrophic shifts in ecosystems. *Nature* 413: 591-596; Lenton, T., H. Held, E. Kriegler, J. W. Hall, W. Lucht, S. Rahmstorf, and H. J. Schellnhuber. 2008. Tipping elements in the Earth's climate system. *Proceedings of the National Academy of Sciences* 105: 1786-1793.

266. Natural Capital Project. Available at: https://www.naturalcapitalproject.org.

267. Goldstein, J. G. Caldarone, T. K. Duarte, D. Ennaanay, N. Hannahs, G. Mendoza, S. Polasky, S. Wolny, and G. C. Daily. 2012. Integrating ecosystem service tradeoffs into land-use decisions. *Proceedings of the National Academy of Sciences* 109(19): 7565-7570.

268. Schenk, R., and P. White, eds. 2014. *Environmental Life Cycle Assessment: Measuring the Environmental Performance of Products*. Vashon Island, WA: American Center for Life Cycle Assessment.

269. Freeman, A. M. III, J. Herriges, and C. L. Kling. 2014. *The Measurement of Environmental and Resource Values: Theory and Methods*, 3rd Ed. New York: Resources for the Future Press.

270. Johnston, R. J., J. Rolfe, R. S. Rosenberger, and R. Brouwer, eds. 2015. *Benefit Transfer of Environmental and Resource Values: A Guide for Researchers and Practitioners*. Dordrecht, The Netherlands: Springer.

271. Elkington, J. 1997. *Cannibals with Forks: The Triple Bottom Line of 21st Century Business*. Oxford, UK: Capstone.

272. U.S. Environmental Protection Agency. 2017. Safer Choice: Design for the Environment: Programs, Initiatives, and Projects.

273. National Research Council. 2014. *Sustainability Concepts in Decision-Making: Tools and Approaches for the U.S. Environmental Protection Agency*. Washington, DC: The National Academies Press.

274. Dilling, L., and M. C. Lemos. 2011. Creating usable science: Opportunities and constraints for climate knowledge use and their implications for science policy. *Global Environmental Change* 21(2): 680-689.

275. Bucchi, M., and B. Trench. 2008. *Handbook of Public Communication of Science and Technology*. Routledge. Available at: https://moodle.ufsc.br/pluginfile.php/1485212/mod_resource/content/1/Handbook-of-Public-Communication-of-Science-and-Technology.pdf [accessed April 2, 2018].

276. Nisbet, M. C., and D. A. Scheufele, 2009. What's next for science communication? Promising directions and lingering distractions. *American Journal of Botany* 96(10): 1767-1778. Available at: http://www.amjbot.org/content/96/10/1767.full.

277. U.S. Census Bureau. 2015. American Community Survey Public Use Microdata Sample; Blaney, L., J. Perlinger, S. Bartelt-Hunt, R. Kandiah, and J. Ducoste. 2017. Another grand challenge: Diversity in environmental engineering. *Environmental Engineering Science* 35(6):568-572.

278. Herring, C. 2009. Does diversity pay?: Race, gender, and the business case for diversity. *American Sociological Review* 74(2): 208-224; Hunt, V., D. Layton, and S. Prince. 2014. *Diversity Matters*. London: McKinsey & Co.

279. Baumol, W. J., and W. E. Oates. 1988. *The Theory of Environmental Policy*, 2nd Ed. Cambridge, UK and New York: Cambridge University Press; Sterner, T. 2003. *Policy Instruments for Environmental and Natural Resource Management*. Washington, DC: Resources for the Future.

280. Weiss, J. A., and M. Tschirhart. 1994. Public information campaigns as policy instruments. *Journal of Policy Analysis and Management* 13(1): 82-119.

281. Allcott, H. 2011. Social norms and energy conservation. *Journal of Public Economics* 95: 1082-1095; Schultz, P. W., J. M. Nolan, R. B. Cialdini, N. J. Goldstein, and V. Griskevicius. 2007. The constructive, destructive, and reconstructive power of social norms. *Psychological Science* 18: 429-434.

282. Larrick, R. P., and J. B. Soll. 2008. Economics. The MPG illusion. *Science* 320: 1593-1594.

283. Thaler, R. H., and C. R. Sunstein. 2008. *Nudge: Improving Decisions about Health, Wealth and Happiness*. New Haven, CT: Yale University Press.

284. Vandenbergh, M. P., P. C. Stern, G. T. Gardner, T. Dietz, and J. M. Gilligan. 2010. Implementing the behavioral wedge: Designing and adopting effective carbon emissions reduction programs. *Environmental Law Reporter* 40: 10547-10554.

285. Johnson E. J., and D. Goldstein. 2003. Do defaults save lives? *Science* 302(5649): 1338-1339.

286. Beshears, J., J. J. Choi, D. Laibson, and B. C. Madrian. 2009. The importance of default options for retirement saving outcomes: Evidence from the United States. Pp. 167-195 in *Social Security Policy in a Changing Environment*. Chicago: University of Chicago Press; Halpern, S. D., P. A. Ubel, and D. A. Asch. 2007. Harnessing the Power Of Default Options To Improve Health Care. *New England Journal of Medicine* 357: 1340-1344; Ebeling, F., and S. Lotz. 2015. Domestic uptake of green energy promoted by opt-out tariffs. *Nature Climate Change* 5(9): 868.

287. Thaler, R. H., and C. R. Sunstein. 2008. Nudge: Improving Decisions about Health, Wealth and Happiness. New Haven, CT: Yale University Press.

288. Waissbein, O., Y. Glemarec, H. Bayraktar, and T. S. Schmidt. 2013. *Derisking Renewable Energy Investment. A Framework to Support Policymakers in Selecting Public Instruments to Promote Renewable Energy Investment In Developing Countries*. New York: United Nations Development Programme.

289. Smith, J. 2014-2015. Sunshine: India's new cash crop. International Water Management Institute.

290. Litke, D. W. 1999. *Review of Phosphorus Control Measures in the United States and Their Effects on Water Quality*. Water-Resources Investigations Report 99-4007. Denver, CO: U.S. Geological Survey.

291. ABET Engineering Accreditation Commission. 2017. Criteria for Accrediting Engineering Programs. Available at: http://www.abet.org/wp-content/uploads/2018/02/E001-18-19-EAC-Criteria-11-29-17.pdf.

292. Department for Professional Engineers. 2014. Professionals in the Workplace: Engineers. Available at: http://dpeaflcio.org/programs-publications/professionals-in-the-workplace/scientists-and-engineers.

293. Examples include Olin College, Dartmouth College, Texas A&M University, the University of Michigan, and Smith College.

294. National Academy of Engineering. 2004. *The Engineer of 2020: Visions of Engineering in the New Century*. Washington, DC: The National Academies Press.

295. Grand Scholars Program. National Academies of Engineering. Available at: http://www.engineeringchallenges.org/GrandChallengeScholarsProgram.aspx.

296. Duderstadt, J. 2009. Engineering for a Changing World, Pp. 17-26 in *Holistic Engineering Education: Beyond Technology*. D. Grasso and M. Burkins, eds. New York: Springer.

297. National Research Council. 2012. *Research Universities and the Future of America: Ten Breakthrough Actions Vital to Our Nation's Prosperity and Security*. Washington, DC: The National Academies Press; President's Council of Advisors on Science and Technology. 2012. *Transformation and Opportunity: The Future of the U.S. Research Enterprise*. Executive Office of the President.

298. National Academy of Sciences, National Academy of Engineering, and Institute of Medicine. 2005. *Facilitating Interdisciplinary Research*. Washington, DC: The National Academies Press; American Academy of Arts & Sciences. 2013. *ARISE 2: Unleashing America's Research & Innovation Enterprise*. Cambridge, MA.
299. National Academy of Sciences, National Academy of Engineering, and Institute of Medicine. 2005. *Facilitating Interdisciplinary Research*. Washington, DC: The National Academies Press.
300. National Research Council. 2014. *Convergence: Facilitating Transdisciplinary Integration of Life Sciences, Physical Sciences, Engineering, and Beyond*. Washington, DC: The National Academies Press; National Research Council. 2015. *Enhancing the Effectiveness of Team Science*. Washington, DC: The National Academies Press.
301. Pollack, M., and M. Snir. 2008. Best Practices Memo: Promotion and Tenure of Interdisciplinary Faculty. Computing Research Association; University of Southern California. 2011. Guidelines for Assigning Authorship and for Attributing Contributions to Research Products and Creative Works.
302. Pittman, J., H. Tiessen, and E. Montaña. 2016. The evolution of interdisciplinarity over 20 years of global change research by the IAI. *Current Opinion in Environmental Sustainability* 19: 87-93.
303. National Academies of Sciences, Engineering, and Medicine. 2017. *A New Vision for Center-Based Engineering Research*. Washington, DC: The National Academies Press.
304. Palmer, M. A., J. G. Kramer, J. Boyd, and D. Hawthorne. 2016. Practices for facilitating interdisciplinary synthetic research: The National Socio-Environmental Synthesis Center (SESYNC). *Current Opinion in Environmental Sustainability* 19: 111-122.

FIGURE SOURCES

Grand Challenge 1

Figure 1-1 FAO Aquastat database, http://www.fao.org/nr/water/aquastat/tables/WorldData-Withdrawal_eng.pdf.
Figure 1-2 Priya Shyamsundar. The Nature Conservancy.
Figure 1-3 iStock/shirnosov.
Figure 1-4 Adapted from High Level Panel of Experts. 2014. Food Losses and Waste in the Context of Sustainable Food Systems. A Report by the High-Level Panel of Experts on Food Security and Nutrition of the Committee on World Food Security. Rome: FAO.
Figure 1-5 Adapted from Organisation for Economic Co-operation and Development. 2012. *OECD Environmental Outlook to 2050: The Consequences of Inaction.*
Figure 1-6 Gassert, F., M. Luck, M. Landis, P. Reig, and T. Shiao. 2015. Aqueduct Global Maps 2.1: Constructing Decision-Relevant Global Water Risk Indicators. Working Paper. Washington, DC: World Resources Institute.
Figure 1-7 Courtesy of Liang Dong, Iowa State University.
Sidebox Figure Pamela Burroff-Murr, Purdue University in Gençer, E., C. Miskin, X. Sun, M. R. Khan, P. Bermel, M. A. Alam, and R. Agrawal. 2017. Directing solar photons to sustainably meet food, energy, and water needs. *Scientific Reports* 7: 3133.
Figure 1-8 National Research Council. 2010. *Electricity from Renewable Resources: Status, Prospects, and Impediments.* Washington, DC: The National Academies Press.

Grand Challenge 2

Figure 2-1 Berkeley Earth. 2018. Global Temperature Report for 2017. Available at: http://berkeleyearth.org/global-temperatures-2017.
Figure 2-2 U.S. Environmental Protection Agency. 2018. *Inventory of U.S. Greenhouse Gas Emissions and Sinks: 1990-2016.* Available at https://www.epa.gov/ghgemissions/inventory-us-greenhouse-gas-emissions-and-sinks-1990-2016.
Figure 2-3 U.S. Global Change Research Program. 2017. Climate Science Special Report: Fourth National Climate Assessment, Vol. 1. D. J. Wuebbles, D. W. Fahey, K. A. Hibbard, D. J. Dokken, B. C. Stewart, and T. K. Maycock, eds. Washington, DC: USGCRP.
Figure 2-4 Gasparrini, A., Y. Guo, F. Sera, A. M. Vicedo-Cabrera, V. Huber, S. Tong, M. de Sousa Zanotti Stagliorio Coelho, P. H. Nascimento Saldiva, E. Lavigne, P. Matus Correa, N. Valdes Ortega, H. Kan, S. Osorio, J. Kyselý, A. Urban, J. J. K. Jaakkola, N. R. I. Ryti, M. Pascal, P. G. Goodman, A. Zeka, P. Michelozzi, M. Scortichini, M. Hashizume, Y. Honda, M. Hurtado-Diaz, J. C. Cruz, X. Seposo, H. Kim, A. Tobias, C. Iñiguez, B. Forsberg, D. O. Åström, M. S. Ragettli, Y. L. Guo, C.-F. Wu, A. Zanobetti, J. Schwartz, M. L. Bell, T. N. Dang, D. D. Van, C. Heaviside, S. Vardoulakis, S. Hajat, A. Haines, and B. Armstrong. 2017. Projections of temperature-related excess mortality under climate change scenarios, *The Lancet Planetary Health* 1(9).

Grand Challenge 3

Box 3-1 Hoornweg, D., and P. Bhada-Tata. 2012. What a Waste: A Global Review of Solid Waste Management. Washington, DC: World Bank.
Figure 3-2 Data from GBD 2016 Risk Factors Collaborators. 2017. Global, regional, and national comparative risk assessment of 84 behavioural, environmental and occupational, and metabolic risks or cluster of risks, 1990-2016: A systematic analysis for the Global Burden of Disease Study 2016. *The Lancet* 390(10100): 1345-1422.
Figure 3-3 Adapted from Procurement Opportunities in the Circular Economy. Anthesis News + Insights [blog]. Available at: https://blog.anthesisgroup.com/procurement-in-circular-economy.

Grand Challenge 4

Figure 4-1 ©Nic Lehoux for the Bullitt Center.
Box 4-2 Sidewalk Labs.

Grand Challenge 5

Box 5-1 Goldstein, J., G. Caldarone, T. K. Duarte, D. Ennaanay, N. Hannahs, G. Mendoza, S. Polasky, S. Wolny, and G. C. Daily. 2012. Integrating ecosystem service tradeoffs into land-use decisions. Proceedings of the National Academy of Sciences 109(19): 7565-7570.
Figure 5-1 Adapted from Moss, R., P. L. Scarlett, M. A. Kenney, H. Kunreuther, R. Lempert, J. Manning, B. K. Williams, J. W. Boyd, E. T. Cloyd, L. Kaatz, and L. Patton. 2014. Decision support: Connecting science, risk perception, and decisions. Pp 620-647 in *Climate Change Impacts in the United States: The Third National Climate Assessment*, J. M. Melillo, T. C. Richmond, and G. W. Yohe, eds. U.S. Global Change Research Program. Available at: https://nca2014.globalchange.gov/report/response-strategies/decision-support.
Figure 5-2 Wikimedia Commons.
Box (on incentivizing water conservation with smart solar pumps): Prashanth Vishwanathan/IMWI.

APPENDIX A

STATEMENT OF TASK

An ad hoc committee of the Water Science and Technology Board of the National Academies of Sciences, Engineering, and Medicine will undertake a study to identify high-priority challenges and opportunities for the broad field of environmental engineering for the next several decades. Given the current and emerging environmental challenges of the 21st century, a study that describes how the field of environmental engineering and its aligned sciences might evolve to better address these needs could serve as a guide to the community and help frame research priorities. These should be significant societal challenges that will require the expertise of environmental engineering and its aligned sciences to resolve or manage. For each challenge, the committee will:

- Discuss the relevance of the challenge, its magnitude, and implications;
- Identify the key questions or issues related to the challenge that require the expertise of environmental engineering to address;
- Discuss the state of knowledge and practice in environmental engineering and aligned sciences relevant to these questions and issues; and
- Identify areas where knowledge and practice need to advance to address these challenges.

APPENDIX B

BIOGRAPHICAL SKETCHES OF COMMITTEE MEMBERS

Domenico Grasso, *Chair*, is chancellor at the University of Michigan–Dearborn. Previously, he was provost of the University of Delaware, Dean of Engineering and Mathematical Sciences, Vice President for Research at the University of Vermont, Founding Director of the Picker Engineering Program at Smith College, and Department Head of Civil and Environmental Engineering at the University of Connecticut. Dr. Grasso has been a Visiting Scholar at the University of California, Berkeley, a NATO Fellow, and an Invited Technical Expert to the United Nations in Vienna, Austria. He is currently editor-in-chief of the journal *Environmental Engineering Science*, and has served as vice chair of the U.S. Environmental Protection Agency Science Advisory Board, and president of the Association of Environmental Engineering & Science Professors. Dr. Grasso's research has focused on the ultimate fate of contaminants in the environment with primary emphasis on colloidal and interfacial processes and environmental chemistry. He has also been active in engineering education reform and views the field of engineering as well poised to serve as a bridge between science and humanity. Dr. Grasso has a B.Sc. from Worcester Polytechnic Institute, an M.S.C.E. from Purdue University, and Ph.D. from the University of Michigan.

Craig H. Benson (NAE) is dean of the School of Engineering and Applied Sciences and the Janet Scott Hamilton and John Downman Hamilton Professor of Civil Engineering in the Department of Civil and Environmental Engineering at the University of Virginia. His research focus areas include engineered barriers for waste containment systems, engineering for sustainability and life-cycle analysis, sustainable infrastructure, and beneficial use of industrial by-products in infrastructure. He has more than 300 research publications and three U.S. patents. Prior to his position at the University of Virginia, Dr. Benson served at the University of Wisconsin–Madison, where he chaired the Department of Civil and Environmental Engineering and the Department of Geological Engineering, co-directed the Office of Sustainability, and served as director of Sustainability Research and Education for the university. Dr. Benson is a member of the National Academy of Engineering, a fellow of ASTM International and the American Society of Civil Engineers, and a Diplomate of Geotechnical Engineering in the Academy of Geo-Professionals. Dr. Benson received a B.S. in civil engineering from Lehigh University and an M.S. and a Ph.D. in civil engineering and geoenvironmental engineering from the University of Texas at Austin.

Amanda Carrico an assistant professor of Environmental Studies at the University of Colorado, Boulder. She is an interdisciplinary environmental social scientist. Her work draws on the fields of psychology (her home discipline), sociology, and economics to examine how individuals make environmentally relevant decisions. Her research focuses on the adoption of behaviors and innovations in response to environmental stress, and the beliefs and perceptions that underpin decision making. She has examined these questions within the context of household and neighborhood decision making in the United States and small-holding agriculture in South Asia. Dr. Carrico received a B.A. from Transylvania University, a Ph.D. in social psychology from Vanderbilt University, and completed a postdoctoral fellowship at the Vanderbilt Institute for Energy and Environment.

Kartik Chandran is a professor in the Department of Earth and Environmental Engineering and the Henry Krumb School of Mines at Columbia University. Dr. Chandran's research focuses on environmental microbiology and biotechnology, reengineering the global nitrogen cycle, sustainable sanitation and wastewater treatment, and microbial platforms for resource recovery. His laboratory employs multidisciplinary strategies to study microbial communities in natural and engineered systems to better understand these communities and their ability to be harnessed for environmental and public health objectives such as waste treatment and improved approaches to clean water, sanitation, and hygiene. Dr. Chandran was awarded a MacArthur fellowship in 2015 for his work on converting pollutants and waste streams to high-value resources. He has a B.S. in chemical engineering from the Indian Institute of Technology and a Ph.D. in environmental engineering from the University of Connecticut.

G. Wayne Clough (NAE) is secretary emeritus of the Smithsonian Institution, and president emeritus of the Georgia Institute of Technology. Dr. Clough served as president of the Georgia Institute of Technology from 1994 to 2008 and as the secretary of the Smithsonian Institution from 2008 to 2014. He previously held faculty appointments at Duke University, Stanford University, and Virginia Polytechnic Institute and State University, where he also served as chair of the Department of Civil and Environmental Engineering and Dean of the College of Engineering. He was provost and vice president of the University of Washington just before coming to Georgia Tech. Dr. Clough's research interests include higher education, civil engineering design and construction, digital learning communities, engineering solutions around climate change, biodiversity conservation, and geotechnical engineering. Dr. Clough earned a B.S.C.E. and M.S.C.E. from the Georgia Institute of Technology and a Ph.D. in geotechnical engineering from the University of California, Berkeley.

John C. Crittenden (NAE) is Hightower Chair and Georgia Research Alliance Eminent Scholar in Environmental Technologies in the School of Civil and Environmental Engineering and director of the Brooks Byers Institute for Sustainable Systems at the Georgia Institute of Technology. Dr. Crittenden's research interests include pollution prevention, physiochemical treatment processes, groundwater transport of organic chemicals, and modeling of water treatment processes. Dr. Crittenden's current research focus is working with other academics and institutions on the challenge of sustainable urban infrastructure systems, including sustainable materials, advanced modeling of urban systems, and sustainable engineering pedagogy. He is a member of the National Academy of Engineering and the Chinese Academy of Engineering. He has a B.S.E. in chemical engineering and an M.S.E. and a Ph.D. in environmental engineering from the University of Michigan, Ann Arbor.

Daniel S. Greenbaum is president and chief executive officer of the Health Effects Institute (HEI). Mr. Greenbaum leads HEI's efforts to provide public and private decision makers—in the United States, Asia, Europe, and Latin America—with high-quality, relevant, and credible science about the health effects of air pollution to inform air quality decisions in the developed and developing world. Mr. Greenbaum has over three decades of governmental and nongovernmental experience in environmental health. Prior to coming to HEI, he served as commissioner of the Massachusetts Department of Environmental Protection, where he was responsible for the Commonwealth's response to the Clean Air Act, as well as its efforts on pollution prevention, water pollution, and solid and hazardous waste. Mr. Greenbaum has been a member of the U.S. National Academies' Board on Environmental Studies and Toxicology and vice chair of its Committee for Air Quality Management in the United States. He served on the Committee on the Hidden Costs of Energy and on their Committee on Science for EPA's Future. In 2010, Mr. Greenbaum received the Thomas W. Zosel Outstanding Individual Achievement Award from the U.S. EPA for his contributions to advancing clean air. Mr. Greenbaum holds bachelor's and master's degrees in city planning from the Massachusetts Institute of Technology.

Steven P. Hamburg is chief scientist at the Environmental Defense Fund (EDF) where he oversees and ensures the scientific integrity of the EDF's positions and programs and facilitates collaborations with researchers from a diversity of institutions and countries. He also helps identify emerging science relevant to EDF's mission. Dr. Hamburg plays a leading role in EDF's research efforts, including work on quantifying methane emissions from the natural gas supply chain and the use of emerging sensor technologies in improving our understanding of air pollution and related impacts on human health. He has been actively involved in biogeochemistry and forest ecology research for more than 35 years, and has published more than 100 scientific papers. Prior to joining EDF, Dr. Hamburg spent 25 years on the faculties at the University of Kansas and Brown University. While at Brown he founded and directed the Global Environment Program at the Watson Institute for International Studies. He also started one of the first university-wide sustainability programs in 1990 at the University of Kansas. Dr. Hamburg has been the recipient of several awards, including recognition by the Intergovernmental Panel on Climate Change as contributing to its award of the 2007 Nobel Peace Prize. He is currently a member of the National Academies' Board on Environmental Studies and Toxicology. He earned a B.A. from Vassar College and an M.S. and a Ph.D. in forest ecology from Yale University.

Thomas C. Harmon is a professor and chair of the Department of Civil & Environmental Engineering and a founding faculty member at the University of California (UC), Merced. Prior to joining the faculty at UC Merced, he served on the faculty of the Department of Civil & Environmental Engineering at the University of California, Los Angeles. Dr. Harmon's research focuses on measuring and modeling flow and transport in natural and engineered systems, including soil, groundwater, and surface water systems. He is the U.S. principal investigator on a Pan-American research project to monitor freshwater ecosystems throughout Central and South America to assess impacts and risks from climate change and local human activities. He has a B.S. in civil engineering from Johns Hopkins University and an M.S. and a Ph.D. in environmental engineering from Stanford University.

James M. Hughes (NAM) is professor emeritus of Medicine (Infectious Diseases), having previously served as professor of Medicine and Public Health with joint appointments in the School of Medicine and the Rollins School of Public Health at Emory University and co-director of the Emory Antibiotic Resistance Center. Prior to joining Emory University in 2005, Dr. Hughes worked at the Centers for Disease Control and Prevention, serving as director of the National Center for Infectious Diseases and as a rear admiral and an assistant surgeon general in the U.S. Public Health Service. Dr. Hughes' research interests focus on emerging and reemerging infectious diseases, antimicrobial resistance, health care–associated infections, vectorborne and zoonotic diseases, foodborne and waterborne diseases, vaccine-preventable diseases, rapid detection of and response to infectious diseases and bioterrorism, and strategies for strengthening public health capacity at the local, national, and global levels. He is a member of the National Academy of Medicine (NAM) and a fellow of the Infectious Diseases Society of America, the American Society of Tropical Medicine and Hygiene, the American Academy of Microbiology, and the American Association for the Advancement of Science. He previously served as president of the Infectious Diseases Society of America. He has served on the Health and Medicine Division/NAM Forum on Microbial Threats from 1996 to 2017 and as vice chair of the Forum from 2009 to 2017. Dr. Hughes received his B.A. and M.D. from Stanford University.

Kimberly L. Jones is professor and chair of the Department of Civil and Environmental Engineering at Howard University and acting associate dean for Research and Graduate Education in the College of Engineering and Architecture. Dr. Jones' research interests include development of membrane processes for environmental applications, physical-chemical processes for water and wastewater treatment, remediation of emerging contaminants, drinking water quality, and environmental nanotechnology. Dr. Jones currently serves on the Science Advisory Board of the U.S. Environmental Protection Agency and as chair of the Drinking Water Committee of the Science Advisory Board. She has served on the Water Science and Technology Board and several committees of the National Academies. She served as the deputy director of the Keck Center for Nanoscale Materials for Molecular Recognition at Howard University. Dr. Jones has received a Top Women in Science award from the National Technical Association, a National Science Foundation CAREER award, and Top Women Achievers award from Essence Magazine. She also served as an associate editor of the *Journal of Environmental Engineering (ASCE)*. She received her B.S. in civil engineering from Howard University, her M.S. in civil and environmental engineering from the University of Illinois, and her Ph.D. in environmental engineering from the Johns Hopkins University.

Linsey C. Marr is the Charles P. Lunsford Professor of Civil and Environmental Engineering at Virginia Polytechnic Institute and State University. Dr. Marr's research interests include characterizing the emissions, fate, and transport of air pollutants in order to provide the scientific basis for improving air quality and health. She also conducts research on the environmental fate of nanomaterials and airborne transmission of infectious diseases. She received the New Innovator Award from the director of the National Institutes of Health in 2013. Dr. Marr received a B.S. degree in engineering science from Harvard University and a Ph.D. degree in civil and environmental engineering from University of California, Berkeley.

Robert Perciasepe is president of the Center for Climate and Energy Solutions, which is widely recognized in the United States and internationally as a leading, independent voice for practical policy and action to address the challenges of energy and climate change. Mr. Perciasepe has been an environmental policy leader in and outside government for more than 30 years, most recently as deputy administrator of the U.S. Environmental Protection Agency (EPA). He is a respected expert on environmental stewardship, natural resource management, and public policy, and has built a reputation for bringing stakeholders together to solve issues. While Mr. Perciasepe served as deputy administrator from 2009 to 2014, EPA set stricter auto emission and mileage standards, increased protections for the nation's streams and rivers, and developed carbon emission standards for power plants. Mr. Perciasepe was previously assistant administrator for both the agency's water and clean air programs, leading efforts to improve the safety of America's drinking water and lower sulfur levels in gasoline to reduce smog. He is a member of the National Academies' Board on Environmental Studies and Toxicology, the National Petroleum Council, and the North American Climate Smart Agriculture Alliance Steering Committee. Mr. Perciasepe holds a master's degree in planning and public administration from Syracuse University and a B.S. in natural resources from Cornell University.

Stephen Polasky (NAS) is the Regents Professor and Fesler-Lampert Professor of Ecological and Environmental Economics at the University of Minnesota, St. Paul. His research focuses on issues at the intersection of ecology and economics and includes the impacts of land use and land management on the provision and value of ecosystem services and natural capital, biodiversity conservation, sustainability, environmental regulation, renewable energy, and common property resources. Dr. Polasky is a member of the National Academy of Sciences, and he is also a fellow of the American Academy of Arts and Sciences, the American Association for the Advancement of Science, and the Association of Environmental and Resource Economists. He has a B.A. from Williams College and a Ph.D. in economics from the University of Michigan.

Maxine L. Savitz (NAE) is a retired general manager of Technology/Partnerships at Honeywell, Inc. (formerly Allied Signal). She is a member and served two terms as vice president of the National Academy of Engineering (2006-2014). Dr. Savitz was appointed to the President's Council of Advisors for Science and Technology in 2009 and served through 2017; she served as vice co-chair (2010-2017). Dr. Savitz was employed at the U.S. Department of Energy (DOE) and its predecessor agencies (1974-1983) and served as the deputy assistant secretary for conservation. Dr. Savitz serves on the advisory bodies for Pacific Northwest National Laboratory and Sandia National Laboratories. She serves on the Massachusetts Institute of Technology visiting committee for sponsored research activities. Past board memberships include the American Council for an Energy Efficient Economy, Jet Propulsion Laboratory, National Science Board, Secretary of Energy Advisory Board, Defense Science Board, Electric Power Research Institute, Draper Laboratories, and the Energy Foundation. Dr. Savitz's awards and honors include elected fellow of the American Academy of Arts & Sciences in 2013; C3E Lifetime Achievement Award in 2013; the Orton Memorial Lecturer Award (American Ceramic Society) in 1998; the DOE Outstanding Service Medal in 1981;

the President's Meritorious Rank Award in 1980; recognition by the Engineering News Record for Contribution to the Construction Industry in 1979 and 1975; and the MERDC Commander Award for Scientific Excellence in 1967. She is the author of about 20 publications. Dr. Savitz has served on numerous National Research Council committees and has participated in multiple National Academies' activities. She is a member of the Division Committee on Engineering and Physical Sciences. Dr. Savitz received a B.A. in chemistry from Bryn Mawr College and a Ph.D. in organic chemistry from the Massachusetts Institute of Technology.

Norman R. Scott (NAE) is professor emeritus in the Department of Biological and Environmental Engineering at Cornell University in the College of Agriculture and Life Sciences (CALS) and College of Engineering. He retired in 2011 after serving the university for over 40 years, dedicating 14 years as director for research for the Cornell University Agricultural Experiment Station and vice president for research and advanced studies. His early research on thermoregulation in animals was crucial in defining the broad set of biological engineering topics that remain important today. His recent research interests include development of sustainable communities with emphasis on renewable energies, including biologically derived fuels, managed ecosystems, and industrial ecology. Dr. Scott is a member of the National Academy of Engineering and served as chair of the National Academies' Board on Agriculture and Natural Resources from 2009 to 2015. Dr. Scott earned a B.S. in agricultural engineering from Washington State University and a Ph.D. from Cornell University.

R. Rhodes Trussell (NAE) is the founder and chairman of Trussell Technologies, Inc., a niche firm focused on process and water quality. Dr. Trussell is an authority on the criteria for water quality and the methods for achieving them. He has worked on the design for numerous water treatment plants, ranging in capacity from less than 1 gallon per minute to nearly 1 billion gallons per day. Dr. Trussell has a special interest in emerging water sources, particularly wastewater reuse, seawater desalination, and recovery of contaminated groundwater. Before founding Trussell Technologies, Inc., he spent 33 years with MWH Global Inc. as it grew from a 50-person California firm to a 6,800-person multinational operating in 40 countries. While at MWH, he rose to become director of Applied Technology and director of Corporate Development as well as a member of both the Board of Directors and the Executive Committee. Dr. Trussell served for more than 10 years on the U.S. Environmental Protection Agency's Science Advisory Board, on 11 committees for the National Academies as well as chair of the Water Science and Technology Board. For the International Water Association, Dr. Trussell served on the Scientific and Technical Council, on two editorial boards, and on the Program Committee for five World Congresses. Dr. Trussell has B.S., M.S., and Ph.D. degrees in environmental engineering from the University of California, Berkeley.

Julie Zimmerman is an internationally recognized engineer whose work is focused on advancing innovations in sustainable technologies. Dr. Zimmerman is jointly appointed as a professor in the Department of Chemical and Environmental Engineering and School of Forestry and Environmental Studies (FES) at Yale University. She also serves as the senior associate dean for Academic Affairs at FES. Her pioneering work established the fundamental framework for her field with her seminal publications on the "Twelve Principles of Green Engineering" in 2003. The manifestation of this framework is taking place in her research group and includes breakthroughs on the integrated biorefinery, designing safer chemicals and materials, novel materials for water purification, and analyses of the water-energy nexus. Prior to coming to Yale University, Dr. Zimmerman was a program manager at the U.S. Environmental Protection Agency where she established the national sustainable design competition, P3 (People, Prosperity, and Planet), which has engaged design teams from hundreds of universities across the United States. Professor Zimmerman is the coauthor of the textbook, *Environmental Engineering: Fundamentals, Sustainability, Design,* that is used in the engineering programs at leading universities. Dr. Zimmerman earned her B.S. from the University of Virginia and her Ph.D. from the University of Michigan jointly from the School of Engineering and the School of Natural Resources and Environment. In addition, Dr. Zimmerman is an associate editor of the journal *Environmental Science & Technology*, and is a member of the Connecticut Academy of Sciences.